Fault-Tolerant Control for Fully Actuated Systems

Donghua Zhou • Miao Cai

Fault-Tolerant Control for Fully Actuated Systems

Series of Fully Actuated System Approach for Control and AI

 Springer

Donghua Zhou
Southeast University
Nanjing, China

Miao Cai
Southeast University
Nanjing, China

ISBN 978-981-95-0690-3   ISBN 978-981-95-0691-0   (eBook)
https://doi.org/10.1007/978-981-95-0691-0

This work was supported in part by the National Natural Science Foundation of China under Grant 62503111, Grant 62527805, Grant U2541220 and Grant 62033008; in part by the Basic Research Program of Jiangsu under Grant BK20251256; and in part by the Fundamental Research Funds for the Central Universities.

This Springer imprint is published by the registered company Springer Nature Singapore Pte Ltd.
The registered company address is: 152 Beach Road, #21-01/04 Gateway East, Singapore 139721, Singapore

If disposing of this product, please recycle the paper.

Preface

The fully actuated system (FAS) theory is transforming the landscape of control science, offering groundbreaking advantages in stability, simplicity, and versatility. This book is a pioneering exploration of how to harness FAS theory to design fault-tolerant control (FTC) systems capable of overcoming component, actuator, and sensor faults, even in the presence of noise and disturbances. By bridging theoretical innovation and practical application, it sets a new standard for FTC research and development.

The book is divided into two parts. Part I covers the FTC design against single faults, which is elaborated in Chaps. 2–6. Part II addresses the FTC design against multiple faults, which is elaborated in Chaps. 7–8. Structured as a comprehensive monograph, this book presents critical topics, including active FTC, robust adaptive designs, self-healing mechanisms, low-power strategies, and finite-time stabilization—all tailored for FAS models. It also addresses cutting-edge solutions for handling input-nonaffine systems and the simultaneous management of actuator and sensor faults. Featuring parameterized observer and controller designs, this work provides highly adaptable and easily implementable frameworks for engineers and researchers alike.

Whether you are a graduate student in control engineering, an academic researcher, or a technical professional in automation systems, this book equips you with the tools to advance your understanding of FAS and FTC. It not only enriches the FAS theory but also introduces innovative approaches that reduce dependency on model accuracy and improve system resilience.

Acknowledgments The authors would like to gratefully acknowledge the generous financial support provided by the National Natural Science Foundation of China under Grant 62503111, Grant 62527805, Grant U2541220 and Grant 62033008; in part by the Basic Research Program of Jiangsu under Grant BK20251256; and in part by the Fundamental Research Funds for the Central Universities.

Nanjing, China   Donghua Zhou
Jan, 2026   Miao Cai

Competing Interests The authors have no competing interests to declare that are relevant to the content of this manuscript.

Contents

Abbreviations

$\mathbb{R}$	Real number space
$\mathbb{R}^n$	n-dimensional Euclidean space
$\mathbb{R}^{m \times n}$	$m \times n$-dimensional real matrix space
$\emptyset$	Empty set
$[n]$	Set $\{1, 2, \cdots, n\}$
$\mathbf{0}$	Zero vector or matrix with a compatible dimension
$\lvert \cdot \rvert$	Cardinality of a set or absolute value of a scalar
I_n	n-dimensional unit matrix
$\lVert \cdot \rVert$	2-norm of a vector or a matrix
$\lVert \cdot \rVert_\infty$	Infinite norm of a vector
A^T	Transpose of a matrix A
$x^{(i)}$	ith derivative of a variable x
$x^{(0 \sim n-1)}$	$[x^\mathsf{T}, \dot{x}^\mathsf{T}, \cdots, x^{(n-1)\mathsf{T}}]^\mathsf{T}$
$A_{0 \sim n-1}$	$[A_0, A_1, \cdots, A_{n-1}]$
x_i	ith element of a vector x
$A_{i,j}$	Element in row i and column j of a matrix A
$\mathscr{L}_g h$	Lie derivative of h with respect to a vector field g
$\exp(\cdot)$	Exponential function
$\det(\cdot)$	Determinant of a matrix
$\mathrm{diag}(\cdot)$	Diagonal matrix
$\Re(\cdot)$	Real part of a complex number
$\lambda_i(\cdot)$	ith eigenvalue of a matrix
$\lambda_{\min}(\cdot)$	Minimum eigenvalue of a matrix
$\lambda_{\max}(\cdot)$	Maximum eigenvalue of a matrix

Acronyms

FAS	Fully actuated system
FTC	Fault-tolerant control
PFTC	Passive fault-tolerant control
AFTC	Active fault-tolerant control
RAFTC	Robust adaptive fault-tolerant control
LPFTC	Low-power fault-tolerant control
FTFTC	Finite-time fault-tolerant control
SHFTC	Self-healing fault-tolerant control
TD	Tracking differentiator
ESO	Extended state observer

Chapter 1
Introduction

1.1 Research Background and Motivation

In the context of gradually increasing automation levels, the scale and complexity of high-end equipment are constantly rising, leading to an increased risk of dynamic system faults. Various faults in practical systems, if not promptly addressed, not only cause economic losses but also may result in casualties and environmental damage [139]. On July 23, 2011, two trains collided on the Chinese Yong-Wen Line, and the cause of the accident was a malfunction in the train signal control system [69]. On May 8, 2018, a Tesla Model X crashed into a barrier in autonomous driving module, and the cause might involve incorrect decisions by the autonomous driving system and operating system failures [97]. On October 29, 2018, and March 10, 2019, the Boeing 737-MAX passenger aircraft crashed in Indonesia and Ethiopia respectively. The reason might involve the incorrect activation of the maneuvering characteristics enhancement module due to failed attitude sensors [120]. On April 20, 2023, the Super Heavy-Starship assembly launched by SpaceX exploded about 4 minutes after ignition and takeoff. The investigation revealed that the failure of the rocket propulsion system led to complete system loss of control [33]. These major accidents highlight the importance of improving the safety and reliability of dynamic systems. Faults have a relatively minor impact in the early stages, usually only affecting control accuracy, and then gradually evolving into catastrophic faults that can lead to major accidents. For example, a stuck fault in an industrial robotic arm could cause the production line to stop, affecting product efficiency and quality. Fault-tolerant control (FTC) is a key technology for enhancing the safety and reliability of modern equipment, capable of maintaining the safe operation of faulty systems and ensuring the basic completion of system tasks. Therefore, the research on FTC has gradually become an important branch of dynamic system control theory and technology.

Most complex systems in practical engineering are nonlinear systems. For instance, the aerodynamic equations of spacecraft attitude control systems are

© The Author(s) 2026
D. Zhou, M. Cai, *Fault-Tolerant Control for Fully Actuated Systems*,
https://doi.org/10.1007/978-981-95-0691-0_1

highly nonlinear, and the relationship between the reaction rate of a continuous stirred tank reactor and factors such as temperature, pressure, and concentration is nonlinear. Nonlinear dynamic systems usually contains general components, e.g., actuators, and sensors, etc. Component and actuator faults will reduce the control performance, and sensor faults will lead to inaccurate state measurement or state estimation. Among them, the FTC design for sensor faults is more challenging. Although early faults often show characteristics such as smallness and intermittency, they usually only affect the completion level of control tasks, but early faults are bound to develop into permanent faults or even catastrophic faults, which may directly lead to the complete divergence of the control system. The purpose of FTC is to ensure the effective completion of system tasks during the evolution of faults through measures such as fault diagnosis and fault compensation, and to minimize the downtime and maintenance of high-end equipment as much as possible. In some harsh working conditions, the evolution process of faults is very rapid. For example, in the deep underground environment, if the synchronous motor of a rotating directional drilling tool that works at high speed fails to lose its magnetic field, and the system does not have an FTC module, the entire drilling system will quickly lose balance and may even lead to the complete damage of the drilling tool. In addition, during the operation of practical systems, there are often unmodeled dynamics, process disturbances, input perturbations, measurement noise, etc. The FTC technology for handling faults cannot be designed independently of these complex situations. For instance, the FTC design for the rotating directional drilling tool needs to consider nonsmooth nonlinear friction forces, regular and random gyroscope drift, sudden temperature and pressure changes, etc. Therefore, modern FTC technology should have the following characteristics:

1. Different types of faults correspond to different FTC approaches.
2. The severity of faults imposes higher and faster requirements on FTC.
3. FTC needs to take into account the processing functions for unmodeled dynamics, disturbances, noise, etc.
4. Faults amplify system nonlinearity, and FTC should have higher nonlinear adaptability and a simpler design structure.

1.2 Fault-Tolerant Control Overview

Before possible faults or after sudden faults, FTC will re-adjust and re-tune the controller based on the specific failure location and characteristics, ensuring the normal operation of the real-time system, or achieving the basic control goals by sacrificing some performance [141]. The origin of FTC is the integrity control structure defined by Niederlinski in 1971 [78]. Later, in 1980, Siljak analyzed the reliable stabilization problem of multiple redundant control systems [89], and in 1993, Patton wrote the first English review article in the field of FTC [80, 81]. Ye Yinzhong et al. published a review article on FTC in the 1980s [112], and Ge Jianhua et al. published a book on FTC technology [40]. Subsequently, the FTC theory and methods for dynamic systems have developed rapidly. The existing FTC technology is mainly divided into two categories: passive fault-tolerant control (PFTC) and active fault-tolerant control (AFTC). The difference lies in that the AFTC has a controller reconstruction while the PFTC does not [52].

1.2.1 Passive Fault-Tolerant Control

Generally, the PFTC mechanism assumes in advance the possible fault locations and characteristics, and designs a robust controller to be insensitive to possible faults. Such PFTC sacrifices some control performance to handle faults, and its framework is shown in Fig. 1.1. Liu et al. used a radial basis function to approximate the unknown nonlinear function. Based on an actuator fault online estimator, they realized PFTC on the basis of $\mathcal{H}_\infty$ performance optimization [64]. Benosman et al. designed PFTC based on Lyapunov stability for additive actuator faults [8]. Khosrowjerdi et al. solved the input signal that can compensate actuator faults through linear matrix inequality methods [54]. Li et al. designed the decentralized fault-tolerant optimal control for nonlinear large-scale systems with actuator faults based on backstepping and fuzzy control [59]. Xie et al. proposed a data-driven PFTC scheme based on policy iteration for actuator faults such as deviation, interruption, and jamming [105]. Zhang et al. studied the FTC problem with simultaneous actuator and sensor faults using the Takagi-Sugeno fuzzy system method [128]. Ma et al. studied the multi-agent fault-tolerant formation control problem connected by air-ground through direct adaptive compensation methods [75]. Colombo et al. proposed a learning-based fault-tolerant tracking controller for the first-order state-space model of a quadrotor aircraft [16]. Basheer et al. proposed an index approaching law for sliding mode surface under the strategy of disturbance attenuation, and designed PFTC for actuator faults [7]. Most of these PFTC rely on prior information of faults, and then perform robust or robust adaptive performance based on standards such as $\mathcal{H}_\infty$ stability or Lyapunov stability. These PFTC structures are special and their specific structures are related to the system type, fault type, etc.

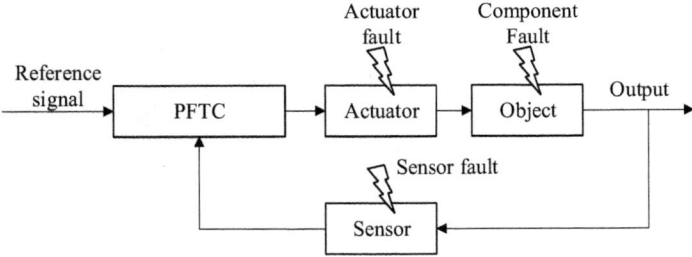

Fig. 1.1 The PFTC structure

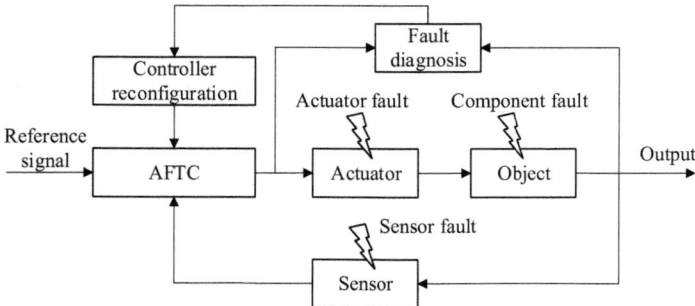

Fig. 1.2 The AFTC structure

1.2.2 Active Fault-Tolerant Control

Parallel to PFTC, AFTC usually requires a fault diagnosis unit. When a fault is detected, the dynamic system reconfigures the corresponding controller to handle the fault. Its structure is shown in Fig. 1.2. Therefore, the performance of AFTC is generally better than that of PFTC. Zhou et al. proposed an online AFTC strategy for sensor faults in a continuous stirred tank reactor based on the Bayesian model and PID state feedback mechanism [140]. Yang et al. designed various controllers for sensor sequence faults in nonlinear systems and then improved the timeliness of the AFTC mechanism under a reasonable switching strategy [110]. Cristofaro et al. used a fault detection observer to estimate the component faults of the nonlinear redundant control system, then utilized an unknown input observer and reset strategy for fault separation, and finally designed a reduced-order controller reconstruction method [17]. Gao et al. studied the AFTC design problem of the satellite attitude control system through the unscented Kalman filter and a PD controller [36]. Chen et al. designed an AFTC framework based on the integral sliding mode control strategy for a class of uncertain nonlinear systems. This closed-loop system can compensate actuator faults and retains the main advantages of integral sliding mode control [12]. Jia et al. regarded the detection delay and estimation error of intermittent faults as new external disturbances, further ensuring that the tracking error can converge to a tiny neighborhood of the origin within a finite time [51]. Sharida et al. proposed a switch fault detection and FTC method for three-phase T-type rectifiers without voltage sensors [87]. Zhao et al. proposed a fault diagnosis and fixed-time sliding mode control strategy based on residual convolutional neural networks and long-short-term memory networks [137]. The AFTC structure is generally more complex than the PFTC structure, and the corresponding engineering design cost is higher. Similarly, the AFTC structure does not have generality, and the fault diagnosis strategies and controller reconstruction rules will change due to different systems and faults.

1.2.3 Summary and Future Directions

In the aforementioned PFTC and AFTC for nonlinear systems, the first-order state-space theory has held an absolute dominant position. However, the controller design corresponding to first-order nonlinear state-space models has the following shortcomings [20]:

1. The original mechanism models of many practical controlled objects are not of the first-order type. For instance, the mechanical arm model given by Lagrange equations is of second-order type. The first-order mathematical model derived from the second-order derivation would mask the full-actuation characteristics and decoupling characteristics of the original system.
2. Most nonlinear fault-tolerant controller structures are complex, and the controller parameters are difficult to tune. The solution process of controllers may also encounter difficulties, such as the differential explosion problem of the backstepping method, etc.
3. The design process of these nonlinear FTC structures is not general, and most controllers can only yield local stability results.

Of course, the first-order state-space theory has advantages in state response analysis, but its advantages in solving control input signals are not obvious. On the one hand, FTC is a research field oriented towards applied control theory or control engineering applications, and direct mechanism models of controlled objects are more suitable for FTC research. On the other hand, FTC technology endows the controller with more comprehensive and complex functions, and the standardized controller structure is more conducive to the addition of fault handling modules and is also beneficial for practical engineering applications. Therefore, it is very necessary to study a standardized fault-tolerant controller that directly targets nonlinear mechanism models [67].

The traditional first-order state-space theory focuses on the analysis of state signals, while the FTC theory focuses on the design of control signals. The two are indirectly related, and the main reason why many FTC technologies based on the first-order state-space theory cannot be practically applied is the complexity of the structural design. In recent years, an emerging control theory—the fully actuated system (FAS) theory [23] is developing rapidly. The FAS structure is parameterized, and the FAS controller is also parameterized, which is more suitable for the theoretical research and technology development of FTC. The FAS theory and the FTC theory are directly related. As a general nonlinear system controllable canonical form, the FAS structure creates more easily designed conditions for control objectives such as FTC and disturbance rejection [42, 68, 127].

1.3   FAS Theory Fundamentals

Inspired by the physically full-actuation characteristics, Professor Duan originally proposed the FAS theory [20, 23]. The physically FASs include mechanical systems, electronic circuit systems, and aerospace systems based on Lagrange equations, Newton's laws of motion, and Kirchhoff's laws, and these original mechanism models are all decoupled in terms of control variables, making it more conducive to the direct design of control input signals [20, 23]. However, physically under-actuated systems represent a wider range of controlled system objects. Therefore, Professor Duan defined a high-order FAS model mathematically and indicated that a large number of physically under-actuated systems can be transformed into mathematically FASs, for example, the under-actuated robot model can be transformed into a mathematically FAS model [28].

1.3.1   Definition and Key Models

Consider the following nonlinear system model:

$$x^{(n)} = F(x^{(0 \sim n-1)}) + G(x^{(0 \sim n-1)})u, \tag{1.1}$$

where n is a positive integer, $x \in \mathbb{R}^m$ is the basic state of the FAS model, $u \in \mathbb{R}^m$ is the control input, $F(x^{(0 \sim n-1)}) \in \mathbb{R}^m$ is the nonlinear dynamic function, and $G(x^{(0 \sim n-1)}) \in \mathbb{R}^{m \times m}$ is the input matrix function.

Definition 1.1 (FAS [20, 23]) Consider $X = [x_1^\mathsf{T}, x_2^\mathsf{T}, \cdots, x_n^\mathsf{T}]^\mathsf{T} \triangleq x^{(0 \sim a-1)} \in \mathbb{R}^{mn}$, if the set $\mathbb{S} \triangleq \{X \mid \det G(X) = 0 \text{ or } \infty, X \in \mathbb{R}^{mn}\}$ satisfies $\mathbb{S} = \emptyset$, then the system (1.1) is called an FAS.

In a physical sense, FAS models can be derived from mechanisms such as Lagrange equations. In a mathematical sense, FAS models are equivalent to many first-order state-space models. In first-order state-space models, controllable linear systems, strict-feedback systems, cascade systems, and linearized-feedback systems, etc., can all be transformed into a high-order FAS structure [23, 24].

Remark 1.1 The core condition of FASs lies in the invertibility of the input matrix $G(x^{(0 \sim n-1)})$. Generally, the input matrices of most controllable systems are invertible. For example, the input matrix of the FAS model from a strict-feedback system is strictly invertible. If there are cases where the input matrix is not invertible, then the sub-FAS theory [32] will provide corresponding stability conditions and controller design methods, while the general state-space theory is difficult to handle such situations.

Lemma 1.1 (Parametric Structure of FASs [20, 23]) *For the FAS model* (1.1), *a parametric controller is selected as*

$$u = -G^{-1}(x^{(0 \sim n-1)}) \left(F(x^{(0 \sim n-1)}) + A_{0 \sim n-1} x^{(0 \sim n-1)} - u_a \right), \qquad (1.2)$$

where $A_{0 \sim n-1} \in \mathbb{R}^{m \times mn}$ is a parametric matrix to be determined, $u_a \in \mathbb{R}^m$ is the feedforward control signal or other auxiliary signals to be determined. Then, a closed-loop linear time-invariant system will be given as

$$\dot{x}^{(0 \sim n-1)} = \Phi_A x^{(0 \sim n-1)} + B_n u_a, \qquad (1.3)$$

where

$$\Phi_A = \begin{bmatrix} 0 & I_m & 0 & \cdots & 0 \\ 0 & 0 & I_m & \cdots & 0 \\ \vdots & & & \ddots & \vdots \\ 0 & 0 & 0 & \cdots & I_m \\ -A_0 & -A_1 & -A_2 & \cdots & -A_{n-1} \end{bmatrix} \in \mathbb{R}^{mn \times mn}, \; B_n = \begin{bmatrix} 0 \\ 0 \\ \vdots \\ 0 \\ I_m \end{bmatrix} \in \mathbb{R}^{mn \times m}.$$

$$(1.4)$$

Lemma 1.2 (Asymptotic Stability of FASs [20, 23]) *If the following Lyapunov equation holds*

$$\Phi_A^{\mathsf{T}} P + P \Phi_A = -\kappa_1 I_{mn}, \qquad (1.5)$$

where κ_1 is a positive constant, and $P \in \mathbb{R}^{mn \times mn}$ is a positive definite solution to the Lyapunov equation (1.5), then the FAS (1.1) under the parametric controller (1.2) is globally asymptotically stable.

The above two lemmas fully demonstrate the advantages of the FAS theory:

1. The modeling and controller design are general and applicable to the analysis, design and synthesis of most control systems.
2. The controller structure is fully parameterized. On one hand, the parametric structure has mathematical standards and solution algorithm procedures. On the other hand, the parametric structure facilitates the optimization of other control performance indicators.
3. The nonlinearity can be decoupled. When the model information is more complete, nonsmooth characteristics, non-integrity characteristics, etc. can also be fully decoupled through FAS controllers.
4. It can apply the mature linear system control theory to the study of nonlinear systems, no longer being limited to the local stability results of nonlinear systems.

Professor Duan also discussed the disturbance rejection and decoupling strategies for FASs [27], as well as the robust adaptive strategies [26] and other structures. Under a certain homeomorphism sense, the FAS is a controllable canonical form of nonlinear systems [28]. When a nonlinear system has state delays, the FAS controller still has corresponding decoupling design [30]. When FASs are generalized to sub-FASs, Professor Duan also defined sub-stability [32]. The sub-FAS approaches have stronger applicability than the general state-space approaches. For example, the general state-space approaches are difficult to stabilize the famous Brockett nonlinear system, while the sub-FAS approaches can guarantee the system stability [31].

1.3.2   State-of-the-Art of FAS

In recent years, the FAS approaches have successfully explored in various fields of control theory and applications:

1. Adaptive control based on the FAS approaches. For nonlinear time-delay systems, the FAS approaches can handle issues such as dead-zone input [131], guaranteed cost control [46], etc. For uncertain high-order strict-feedback systems, Meng et al. combined the FAS approach to achieve event-trigered control [76], Li et al. combined the FAS approach to achieve adaptive iterative learning control [61], Xiao et al. combined the FAS approach to achieve adaptive dynamic surface control [104], and Yan et al. combined the FAS approach and reconstructed filter to achieve adaptive estimation of unknown parameters [108]. In addition, for a class of nonlinear cascade systems, Wang et al. designed a fixed-threshold event-triggered adaptive controller based on the FAS approach [100].
2. Trajectory tracking control based on the FAS approaches. Tian et al. designed various FAS tracking control methods for uncertain robotic arms under various physical constraints, ensuring high-precision tracking performance [92, 93]. Gao et al. studied the robust model tracking control problem for a class of high-order generalized systems [37]. Hu et al. proposed a robust switching adaptive tracking controller for the FAS model, making the tracking error converge to a tiny neighborhood of the origin [47].
3. Predictive control based on the FAS approaches. Liu et al. established a multi-agent FAS model and realized coordinated control through multi-step prediction [63], and further optimized the predictive controller [126]. Considering the communication delay of multi-agents, Zhang et al. proposed a PI-type predictive FAS controller [127]. When facing network attacks, the FAS approach can also enhance the robustness of multi-agent systems [123].
4. Furthermore, the FAS approaches has conducted preliminary explorations in various theoretical fields such as disturbance decoupling control [96, 101], prescribed performance control [130], and impulsive system control [49], and have been successfully applied to practical control systems such as spacecraft [14, 72, 132], unmanned aerial vehicles [70, 71], robots [1, 109], DC microgrids [118, 119], and servo motors [57].

Table 1.1 FAS models and state-space models

State-space models	FAS models
A large number of low-order equations	A small number of high-order equations
Indirectly derived from original systems	Directly derived from original systems
Focusing on the state and beneficial for state response analysis	Focusing on the input and beneficial for controller design

1.3.3 FAS Theory and State-Space Theory

The FAS theory is not completely abandoning state-space models. State-space models have a complete system in linear system analysis, design and synthesis. The FAS theory is based on the mature linear control system theory, so the two control system methodologies are complementary to each other [29]. The comparison of the two dynamic system models is shown in Table 1.1. When the complex characteristics are prominent, the FAS theory has a broader development space.

1.4 Challenges and Opportunities in Fault-Tolerant Control

Since the FAS theory is still in its early stages of research, most existing FAS approaches consider fault-free systems. When the components, actuators and sensors of nonlinear systems fail, the ideal FAS controller is no longer able to provide a completely decoupled linear time-invariant system. The damage to the parametric FAS structure caused by these faults mainly includes the following situations:

1. The FAS model with component faults is

$$x^{(n)} = F(x^{(0\sim n-1)}) + G(x^{(0\sim n-1)})u + \delta, \qquad (1.6)$$

where δ represents possible component faults. When the controller has an ideal parametric structure as given by (1.2), the closed-loop system is

$$x^{(n)} + A_{0\sim n-1}x^{(0\sim n-1)} = u_a + \delta. \qquad (1.7)$$

The closed-loop system (1.7) still has unknown faulty components. This parametric structure has not completed the decoupling design. At this point, the auxiliary signal u_a requires additional analysis and design.

2. The FAS model with actuator faults is

$$x^{(n)} = F(x^{(0\sim n-1)}) + G(x^{(0\sim n-1)})\left[(1-\rho)u + u_f\right], \qquad (1.8)$$

where ρ represents possible actuator drift faults, and u_f represents possible actuator deviation faults. When the controller has an ideal parametric structure as given by (1.2), the closed-loop system is

$$x^n + (1 - \rho)A_{0 \sim n-1}x^{(0 \sim n-1)} = \rho F(x^{(0 \sim n-1)}) + (1 - \rho)u_a + G(x^{(0 \sim n-1)})u_f. \tag{1.9}$$

The nonlinear cancellation principle of the closed-loop system (1.9) is completely ineffective. At this point, the entire FAS controller structure needs to be redesigned.

3. The FAS model with sensor faults is

$$\begin{cases} x^{(n)} = F(x^{(0 \sim n-1)}) + G(x^{(0 \sim n-1)})u, \\ y = x^{(0 \sim n-1)} + s, \end{cases} \tag{1.10}$$

where s represents possible sensor faults. In this case, the controller structure would be

$$u = -G^{-1}(y)\left(F(y) + A_{0 \sim n-1}y - u_a\right), \tag{1.11}$$

and the closed-loop system is

$$x^{(n)} = \check{F}(x^{(0 \sim n-1)}, s, u_a), \tag{1.12}$$

where $\check{F}(\cdot)$ is an extremely complex unknown nonlinear function. The parameterized linear structure has been severely disrupted. At this point, the full-order state observation and parameterized controller need to be redesigned. The challenges posed by sensor faults to the FAS theory are the greatest.

The FTC is the research on applied control theory and technology, and the FAS models are directly derived from the physical mechanisms of original controlled objects. Therefore, the two are closely related. Different types of faults have different degrees of damage to the FAS theory, and the corresponding FTC ideas are also different. Up to now, the FAS theory lacks the FTC module. In order to standardize the design structure of nonlinear FTC, this monograph conducts research on the FTC for high-order FASs.

1.5 Book Structure Overview

1.5.1 The FAS Approach Book Series

This book is part of the following FAS Approach Book Series:
Fully Actuated System Approach for Control and AI

which is to be published with Springer (see http://www.fasta.org.cn for the proposal). The executing period of this book series is from 2025 to 2030. It is affiliated with the following FASTA conference which has been held four times with FASTA 2025 attracting more than 1040 participants:

Title: *International Conference on Fully Actuated System Theory and Applications*

Websites: http://fasta2024.fasta.org.cn; http://fasta2025.scimeeting.cn

As planned, each year within the executing period there will be 4 to 6 books on this topic published. The following titles are planned to appear in 2025:

1. Fully Actuated System Approach,
   Volume I. Global Fully Actuated Systems
   Volume II. Sub-Fully Actuated Systems
   Volume III. Unidirectionally Connected FASs
   by Guang-Ren Duan, Harbin Institute of Technology;
2. Predictive Cooperative Control for High-Order Fully Actuated Multiagent Systems,
   by Da-Wei Zhang, Guo-Ping Liu, Southern University of Science and Technology;
3. Fault-Tolerant Control for Fully Actuated Systems,
   by Dong-Hua Zhou, Miao Cai, Southeast University;
4. Adaptive Constrained Control for High-Order Fully Actuated Nonlinear Systems,
   by Chang-Chun Hua, Liu-Liu Zhang, Peng-Ju Ning, Yanshan University;
5. Cooperative Safety Control of Multiagent Systems—A Fully Actuated System Approach,
   by Ke Zhang, Bin Jiang, Yong-Hao Ma, Yu-Hang Xu, Yuan Lu, Nanjing University of Aeronautics and Astronautics;
6. Modeling and Control of Renewable Power Systems—A Fully Actuated System Approach,
   by Yi Yu, The Hong Kong Polytechnic University,
   Guo-Ping Liu, Southern University of Science and Technology,
   Josep M. Guerrero, Zhejiang University.

1.5.2 About the Monograph

This monograph aims to study the parameterized FTC design of FASs under the conditions of component faults, actuator faults and sensor faults. Various types of faults in dynamic systems will disrupt the structural characteristics and stability design advantages of the FAS theory. More seriously, when uncertainties, process disturbances, input perturbations, measurement noise and other conditions occur in nonlinear systems with faults, the decoupled FAS controller structure requires further design. In response to the key challenges brought by different fault types, different performance requirements, and different system forms to the parameterized controller structures, this monograph proposes corresponding FTC design methods.

The monograph consists of 9 chapters. Chapter 1 introduces the background and significance of the topic, the concepts and research status of FTC and FAS theory, the challenges of faults to the FAS theory, and the structure arrangement of the monograph. Chapters 2 to 8 conduct detailed research on different FTC problems of FASs. Chapter 9 summarizes the main work and looks forward to the future research directions.

Chapter 2 proposes an AFTC framework based on dynamic data modeling for uncertain high-order FASs with component faults, process disturbances, and unmodeled dynamics. By using the Lie derivative technique, the original system model is transformed into a high-order FAS model with unknown nonlinear functions, and the system state is not completely measurable. A unified structure of the observer and controller is established, and the dynamic data modeling is given according to this unified structure. Since this data model can reveal the dynamic characteristics of the closed-loop FAS including fault information, fault diagnosis design and AFTC design are proposed. Compared with the traditional PFTC, the proposed AFTC framework has better fault-tolerant performance for large-amplitude component faults.

Chapter 3 proposes a robust adaptive fault-tolerant control (RAFTC) framework for FASs with actuator faults and process disturbances. Starting from the multi-input multi-output high-order FASs, an extended state observer is designed to estimate the actual high-order tracking error. The parameters of this observer can weaken the peak phenomenon. For multiplicative actuator faults, a robust FTC strategy is adopted, and for additive actuator faults, an adaptive FTC strategy is utilized. Combined with the observer-based FAS controller, the ultimate error can be made as small as possible. The proposed RAFTC technology is more effective than the traditional linear active disturbance rejection control technology.

Chapter 4 proposes a low-power fault-tolerant control (LPFTC) framework for uncertain time-varying FASs with actuator faults, process disturbances, unmodeled dynamics, and high-frequency measurement noise. Under certain mathematical constraints, the FTC idea for actuator faults is similar to that for component faults in Chap. 2. Firstly, the standard FTC framework is given based on the idea in Chap. 2, which can yield bounded stability of the error system, but is very sensitive to high-frequency noise and the corresponding controller signal oscillation phenomenon is obvious. Different from the standard FTC framework, Chap. 4 designs an LPFTC framework based on adjacent error feedback. This LPFTC framework still retains the parameterization structure advantage of FASs. It is mathematically proved that the noise suppression performance of the linear LPFTC framework is superior to that of the standard FTC framework.

Chapter 5 further considers the finite-time convergence performance of the error system. In the general asymptotic stability design, the early FAS controllers stand out due to their parameterized design standard process. However, in high-performance control systems, asymptotic stability design may not meet the corresponding control tasks. To further promote the FAS theory, Chap. 5 redesigns a finite-time fault-tolerant control (FTFTC) framework for FASs based on the principle of homogeneity. The parameterization property is still retained, and the FTFTC can stabilize nonlinear systems faster and compensate actuator faults faster.

Chapter 6 differs from the previous chapters. Component and actuator faults can be estimated and compensated, while sensor faults are difficult to estimate and cannot be compensated, because completely incorrect measurements cannot provide correct state information. Inspired by the physical hardware redundancy, Chap. 6 mathematically defines the observability and redundant observability of FASs and gives an ideal sensor fault tolerance definition, and derives a necessary and sufficient condition between ideal sensor fault tolerance and system redundant observability. Finally, based on dynamic fused observation, a self-healing fault-tolerant control (SHFTC) framework is proposed, which can handle sensor faults in both steady-state and transient processes.

Chapter 7 establishes three AFTC frameworks for high-order FASs with actuator faults, sensor faults, and general measurement noise. The first AFTC framework is based on the research in Chaps. 2 and 6. Although this design can handle both actuator and sensor faults simultaneously, it is significantly affected by measurement noise. This is because high-order systems are highly sensitive to measurement noise, and the high-order controller integrating observations is even more sensitive to measurement noise. To further improve the noise suppression performance, two frameworks are proposed—AFTC saturation design and AFTC dead-zone design. These two frameworks suppress the error feedback term via saturation and dead zone, respectively. Under a special linear FAS structure, the noise suppression performance of saturation design and dead-zone design has been proven to be superior to that of general design. The difference is that AFTC saturation design has a better suppression effect on impulsive-type or intermittent-type noise, while AFTC dead-zone design has a better suppression effect on continuous-type noise.

Chapter 8 extends the input-affine FASs to input-nonaffine FASs, and studies the FTC design problem for input-nonaffine FASs with unknown functions. Under the nonlinear cancellation in FASs, the nonaffine controllers require the inverse function of nonlinear dynamics. The inversions of many nonlinear functions are difficult to be analytically given, and it is even impossible when the nonlinear dynamics are unknown. Starting from the final result of nonlinear cancellation, Chap. 8 establishes a control input dynamic equation, which can make the entire closed-loop system converge to the ideal closed-loop FAS. Additionally, for input-nonaffine FASs with both actuator and sensor faults, a robust FTC structure is designed based on an auxiliary system to ensure the bounded convergence of the error system.

Part I
FTC for FASs Against Single Faults

Chapter 2
Active FTC for FASs Against Component Faults

This chapter introduces an active fault-tolerant control (AFTC) framework tailored for high-order fully actuated systems (FASs) with parameter uncertainties. Beginning with the formulation of nonlinear measurement-based uncertain faulty FAS models, the benefits of this representation are systematically demonstrated through Lie derivative analysis. An integrated FAS adaptive observer-controller mechanism is constructed, from which a closed-loop dynamic data modeling framework is subsequently developed. This structure uncovers the dynamic characteristics including fault features, and naturally yields a fault detectability condition. Leveraging the fault diagnosis model, an AFTC methodology is formulated, thereby extending the theoretical foundation of FASs. The uniform bounded stability is rigorously validated through mathematical proof and experimentally confirmed via case studies.

2.1 Overview of Active Fault-Tolerant Control

The early FAS theory focuses on the analysis, design and synthesis of ideal systems, without considering the real-time safety requirements of dynamic systems. Dynamical systems working in complex environments for a long time will inevitably face various component faults, which will destroy the nonlinear cancellation principle of the FAS theory. There are a great deal of theoretical results on the fault-tolerant control (FTC) design based on the first-order nonlinear state space theory. For example, a $\mathscr{H}_\infty$ dynamic compensator technique is designed to tolerate system faults of uncertain generalized systems [62], a sliding mode control method compensates for boiler faults of heat recovery steam generators [2], a backstepping technique adaptively estimates partial stuck faults [125], a high-performance FTC strategy implements early fault diagnosis, etc. However, the research of FTC under the nonlinear high-order FAS models is still in the preliminary exploration stage.

© The Author(s) 2026
D. Zhou, M. Cai, *Fault-Tolerant Control for Fully Actuated Systems*,
https://doi.org/10.1007/978-981-95-0691-0_2

The FAS theory not only lacks perfect FTC modules but also lacks uncertainty processing modules. Most existing FAS approaches require accurate dynamic system models and full-order state measurements, but strong uncertainties in complex environments can lead to completely unknown system models. When the system dynamics are unknown, the parameterized structures of FASs need to be redesigned. Due to the cost of sensors and other reasons, the high-order system state can not be precisely given. Therefore, it is urgent to study the FTC theory of a class of FAS models whose parameters are completely unknown and high-order states are unmeasurable.

In this chapter, an AFTC framework for uncertain high-order FASs is presented. Firstly, from an uncertain high-order FAS model with nonlinear measurements, the standard FAS system model can be determined via Lie derivative techniques. Then, the observer and controller integrated framework of the integrated FAS model is established, and the dynamic data modeling structure of the closed-loop FAS model is derived. This data structure can reveal the dynamic characteristics of the closed-loop system, including the fault characteristics. Finally, the fault detection condition and the switching strategy of the fault-tolerant controller are given.

2.2 Problem Formulation and Key Preliminaries

Consider a class of uncertain high-order FAS models:

$$
\begin{cases}
x_o^{(n)} = F(x_o^{(0\sim n-1)}, \xi) + G_o(x_o^{(0\sim n-1)}, \xi)u + K_o(x_o^{(0\sim n-1)}, \xi)d, \\
y = H_o(x_o),
\end{cases} \tag{2.1}
$$

where $x_o = [x_{o1}, x_{o2}, \cdots, x_{om}]^\mathsf{T} \in \mathbb{R}^m$ is the basic state, $u \in \mathbb{R}^m$ is the control input, $d \in \mathbb{R}^m$ is the time-varying process disturbance, $y \in \mathbb{R}^m$ is the measurement output, $\xi \in \mathbb{R}$ is the unmodeled dynamics. $F_o(\cdot) = [F_{o1}(\cdot), F_{o2}(\cdot), \cdots, F_{om}(\cdot)]^\mathsf{T} \in \mathbb{R}^m, G_o(\cdot) \in \mathbb{R}, K_o(\cdot) \in \mathbb{R}, H_o(\cdot) = [H_{o1}(\cdot), H_{o2}(\cdot), \cdots, H_{om}(\cdot)]^\mathsf{T} \in \mathbb{R}^m$ are unknown smooth nonlinear functions. $G_o(\cdot)$ meets the full-actuation condition, and the nonlinear FAS (2.1) is controllable and observable.

Define the following vector fields

$$
\begin{cases}
\mathscr{F}_j = [\dot{x}_{oj}, \ddot{x}_{oj}, \cdots, x_{oj}^{(n-1)}, F_{oj}]^\mathsf{T}, j \in [m], \\
\mathscr{G} = [0_{n-1}^\mathsf{T}, G_o(x_o^{(0\sim n-1)}, \xi)]^\mathsf{T}, \\
\mathscr{K} = [0_{n-1}^\mathsf{T}, K_o(x_o^{(0\sim n-1)}, \xi)]^\mathsf{T},
\end{cases} \tag{2.2}
$$

and there are some calculations

$$
\mathscr{L}_\mathscr{G} \mathscr{L}_{\mathscr{F}_j}^r H_{oj} = 0, r \in [n-2], \quad \mathscr{L}_\mathscr{G} \mathscr{L}_{\mathscr{F}_j}^{n-1} H_{oj} \neq 0. \tag{2.3}
$$

The relative order of the nonlinear high-order FAS model (2.1) is n, and the system model (2.1) can be transformed into

$$
\begin{cases}
x^{(n)} = F(x^{(0\sim n-1)}, \xi) + G(x^{(0\sim n-1)}, \xi)u + K(x^{(0\sim n-1)}, \xi)d, \\
y = x,
\end{cases}
\tag{2.4}
$$

where $x \in \mathbb{R}^m$ is the system state after transformation, and $F(\cdot), G(\cdot), K(\cdot)$ are the smooth nonlinear functions after transformation. The special structure of the FAS model makes the above transformation inevitable, and its full-actuation condition is still valid.

Assumption 2.1 The partial derivatives of nonlinear functions $F(\cdot), K(\cdot)$ are norm-bounded, and the disturbance function $D(t) = K(\cdot)d(t)$ satisfies $\sup_{t \geq 0} \|d(t)\| < \infty$, $\sup_{t \geq 0} \|\dot{d}(t)\| < \infty$ and $\sup_{t \geq 0} \|\dot{\xi}(t)\| < \infty$.

Assumption 2.2 For the unknown matrix function $G(\cdot)$, there exists a known nominal matrix $\bar{G}(t) = g(t)I_m, g(t) \in \mathbb{R}, g(t) \neq 0$ satisfying

$$
\sup_{t \geq 0} \left\| \frac{\mathscr{L}_g \mathscr{L}_{\mathscr{F}}^{n-1} H_{oj}}{g(t)} \right\| < \infty, \quad \sup_{t \geq 0} \left\| \frac{d}{dt} \frac{\mathscr{L}_g \mathscr{L}_{\mathscr{F}}^{n-1} H_{oj}}{g(t)} \right\| < \infty.
\tag{2.5}
$$

Remark 2.1 Assumption 2.1 is a common assumption of nonlinear systems, and the specific physical meaning is that the nonlinear function is Lipschitz, the perturbation and its derivatives are norm-bounded [58, 77, 124, 133]. Assumption 2.2 is the basic condition of nonlinear observer design, a completely determined input matrix function or an input matrix function with only a small perturbation meets this requirement [125, 133].

In complex and harsh working environments, dynamic systems will encounter various component faults. Due to the unknown dynamics, the faulty FAS model can be directly described as

$$
\begin{cases}
x^{(n)} = F(x^{(0\sim n-1)}, \xi) + G(x^{(0\sim n-1)}, \xi)u + K(x^{(0\sim n-1)}, \xi)d + \delta, \\
y = x,
\end{cases}
\tag{2.6}
$$

where $\delta \in \mathbb{R}^m$ is the possible component fault, which satisfies $\sup_{t \geq 0} \|\dot{\delta}(t)\| \leq \delta_m < \infty$ and $\sup_{t \geq 0} \|\delta(t)\| < \infty$ with a positive scalar δ_m.

Assuming that the dynamics functions, high-order states, and unmodeled dynamics are known, the completely ideal FAS controller is

$$
u = -G^{-1}(x^{(0\sim n-1)}, \xi)\left(F(x^{(0\sim n-1)}, \xi) + A_{0\sim n-1}x^{(0\sim n-1)} - u_a \right),
\tag{2.7}
$$

and the corresponding closed-loop system is

$$
x^{(n)} + A_{0\sim n-1}x^{(0\sim n-1)} = u_a + K(x^{(0\sim n-1)}, \xi)d + \delta,
\tag{2.8}
$$

where u_a is the auxiliary signal to realize disturbance rejection and fault tolerance. However, the dynamics equations and high-order states of FASs are unknown, so an FAS controller with weak model requirements need to be studied. Define $X = [x_1^\mathsf{T}, x_2^\mathsf{T}, \cdots, x_n^\mathsf{T}]^\mathsf{T} \triangleq x^{(0 \sim n-1)}$, and the total uncertainty is $\varpi = F(X, \xi) + \left(G(X, \xi) - \bar{G}\right)u + K(X, \xi)d + \delta$. Naturally, the FAS controller can be redesigned as

$$u = -\bar{G}^{-1}\left(\varpi + A_{0 \sim n-1}X - u_a\right). \tag{2.9}$$

The controller contains only known parameters, designed parameters and estimated variables, and the physical mechanism is parametric state feedback. The control task of this chapter is to design a fault-tolerant controller for the FAS model (2.6) so that the output signal y tracks a smooth reference signal y_d. The reference signal satisfies $\sup_{t \geq 0} \|y_d^{(0 \sim n)}\| < \infty$. To simplify the derivative calculations of y_d, a tracking differentiator (TD) is designed as

$$\begin{cases} \dot{x}_{Ri} = x_{R(i+1)}, i \in [n], \\ \dot{x}_{R(n+1)} = R^n \varphi\left(x_{R1} - y_d, \frac{x_{R2}}{R}, \cdots, \frac{x_{R(n+1)}}{R^n}\right), \end{cases} \tag{2.10}$$

where $X_R = [x_{R1}^\mathsf{T}, x_{R2}^\mathsf{T}, \cdots, x_{Rn}^\mathsf{T}, x_{R(n+1)}^\mathsf{T}]^\mathsf{T}$ is the TD state, R is the convergence parameter, and $\varphi(\cdot)$ is a Lipschitz function passing through the origin. If $y_d = 0_m$, $R = 1$, the TD (2.10) is globally asymptotically stable.

Lemma 2.1 ([44]) *For any initial value of a TD system, if $R \to \infty$, then x_{R1} converges uniformly to y_d. In the weakly convergent sense, for $2 \leq i \leq n+1$, the TD state x_{Ri} will converge to the corresponding derivative of the reference signal.*

2.3 Dynamic Modeling for Closed-Loop Systems

2.3.1 Observer and Controller Integration

For the FAS model (2.6), an observer and controller integration framework is designed as

$$\begin{cases} \dot{\hat{x}}_i = \hat{x}_{i+1} + \Gamma^{n-i}(t)h_i(\vartheta), i \in [n-1], \\ \dot{\hat{x}}_n = \hat{\varpi} + h_n(\vartheta) + \bar{G}u, \\ \dot{\hat{\varpi}} = \dfrac{1}{\Gamma(t)}h_{n+1}(\vartheta), \\ u = -\bar{G}^{-1}\left[\hat{\varpi} - x_{R(n+1)} + A_{0 \sim n-1}(\hat{X} - X_R)\right], \end{cases} \tag{2.11}$$

where $\vartheta = (y - \hat{x}_1)/\Gamma^n(t)$, $\hat{X} = [\hat{x}_1^\mathsf{T}, \hat{x}_2^\mathsf{T}, \cdots, \hat{x}_n^\mathsf{T}]^\mathsf{T}$ is the state observation, and $\hat{\varpi}$ is the estimation of the total uncertainty ϖ. $h_i(\cdot) \in \mathbb{R}^m$, $i \in [n+1]$ are nonlinear functions to be designed, which satisfy $\|h_i(r)\| \leq \mathcal{H}_i\|r\|, \forall r \in \mathbb{R}^m$ with positive scalars $\mathcal{H}_i, i \in [n+1]$. The time-varying gain $\Gamma(t)$ satisfies

$$\dot{\Gamma}(t) = \begin{cases} -\mathfrak{a}\Gamma(t), \Gamma(t) > \epsilon, \\ 0, \Gamma(t) \leq \epsilon, \end{cases} \quad (2.12)$$

where $\mathfrak{a} > 0$, $\Gamma(0) = 1$, and $\epsilon \in (0, 1)$.

Assumption 2.3 There exist positive scalars $\alpha_1, \alpha_2, \alpha_3, \alpha_4, \beta, \gamma$ and positive definite functions $V_O(\mathfrak{z}), W_O(\mathfrak{z}) : \mathbb{R}^{m(n+1)} \to \mathbb{R}, \forall \mathfrak{z} = [\mathfrak{z}_1^\mathsf{T}, \mathfrak{z}_2^\mathsf{T}, \cdots, \mathfrak{z}_{n+1}^\mathsf{T}]^\mathsf{T}$ satisfying

1. $\alpha_1\|\mathfrak{z}\|^2 \leq V_O(\mathfrak{z}) \leq \alpha_2\|\mathfrak{z}\|^2, \alpha_3\|\mathfrak{z}\|^2 \leq W_O(\mathfrak{z}) \leq \alpha_4\|\mathfrak{z}\|^2$.
2. $\sum\limits_{i=1}^{n}[\frac{\partial V_O}{\partial \mathfrak{z}_i}(\mathfrak{z})]^\mathsf{T} (\mathfrak{z}_{i+1} - h_i(\mathfrak{z}_1)) - [\frac{\partial V_O}{\partial \mathfrak{z}_{n+1}}(\mathfrak{z})]^\mathsf{T} h_{n+1}(\mathfrak{z}_1) \leq -W_O(\mathfrak{z}), \left\|\frac{\partial V_O}{\partial \mathfrak{z}_{n+1}}(\mathfrak{z})\right\| \leq \beta\|\mathfrak{z}\|$.
3. $\sum\limits_{i=1}^{n}[\frac{\partial V_O}{\partial \mathfrak{z}_i}(\mathfrak{z})]^\mathsf{T}\mathfrak{z}_i \leq \gamma W_O(\mathfrak{z})$.

Besides, some parameters satisfy $\|\bar{G}^{-1}[G(X, \xi) - \bar{G}]\|\mathcal{H}_{n+1} < \alpha_3/\beta$.

Remark 2.2 The first two conditions of Assumption 2.3 are common conditions for error convergence of nonlinear observers [45, 53, 84], and the third condition is the basic convergence requirement of nonlinear observers with time-varying gain.

Theorem 2.1 *For the faulty FAS model* (2.6) *under the observer and controller integration* (2.11), *if Assumptions 2.1–2.3 hold, there exist positive scalars* $t_{1\epsilon}, t_{2\epsilon}, t_{R\epsilon}, \Theta_1, \Theta_2, \Theta_3$ *such that*

1. The observation error and tracking error are ultimately uniformly bounded.
2. For any $\epsilon \in (0, \epsilon_0)$, ϵ_0 *is defined in* (2.33), *it holds that*

$$\begin{cases} \|x_i - \hat{x}_i\| \leq \epsilon^{n+2-i}\Theta_1, t \geq t_{1\epsilon}, i \in [n], \\ \|\varpi - \hat{\varpi}\| \leq \epsilon\Theta_1, t \geq t_{1\epsilon} \\ \|x_i - x_{Ri}\| \leq \epsilon\Theta_2, t \geq t_{2\epsilon}, i \in [n], \\ \|y - y_d\| \leq \epsilon\Theta_3, t \geq t_{R\epsilon}. \end{cases} \quad (2.13)$$

Proof The concrete form of the time-varying gain function is

$$\Gamma(t) = \begin{cases} \exp(-\mathfrak{a}t), 0 \leq t < -\frac{\ln\epsilon}{\mathfrak{a}}, \\ \epsilon, t \geq -\frac{\ln\epsilon}{\mathfrak{a}}, \end{cases} \quad (2.14)$$

and $\Gamma(t) \in [\epsilon, 1]$. Define the observation error

$$e_i = \frac{1}{\Gamma^{n+1-i}(t)}(x_i - \hat{x}_i), i \in [n], e_{n+1} = \varpi - \hat{\varpi}, \quad (2.15)$$

the time-dependent derivative of $e_i, i \in [n]$ is

$$
\begin{aligned}
\dot{e}_i &= \frac{x_{i+1} - \hat{x}_{i+1} - \Gamma^{n-i}(t)h_i(\vartheta)}{\Gamma^{n+1-i}(t)} - \frac{(n+1-i)\dot{\Gamma}(t)(x_i - \hat{x}_i)}{\Gamma^{n+2-i}(t)} \\
&= \frac{e_{i+1} - h_i(e_1)}{\Gamma(t)} - \frac{(n+1-i)\dot{\Gamma}(t)e_i}{\Gamma(t)}.
\end{aligned}
\tag{2.16}
$$

The jth element of the derivative of e_{n+1} is

$$
\begin{aligned}
\dot{e}_{n+1,j} &= \dot{\varpi}_j - \frac{1}{\Gamma(t)}h_{n+1,j}(\vartheta) \\
&= K_j \dot{d}_j + \frac{\partial F_j}{\partial \xi}\dot{\xi} + \frac{\partial K_j}{\partial \xi}\dot{\xi}d_j + \dot{\delta}_j + \sum_{i=1}^{n-1}[\frac{\partial F_j}{\partial x_i} + \frac{\partial K_j}{\partial x_i}d_j]^\mathsf{T} x_{i+1} \\
&\quad + [\frac{\partial F_j}{\partial x_n} + \frac{\partial K_j}{\partial x_n}d_j]^\mathsf{T} \varpi \\
&\quad + [\frac{\partial F_j}{\partial x_n} + \frac{\partial K_j}{\partial x_n}d_j]^\mathsf{T}\left[x_{R(n+1)} - \hat{\varpi} - A_{0\sim n-1}(\hat{X} - X_R)\right] - u_j \frac{\mathrm{d}}{\mathrm{d}t}\frac{G_j}{g} \\
&\quad + \frac{G_j - g}{g}\left[\dot{x}_{R(n+1),j} - \frac{h_{n+1,j}(e_1)}{\Gamma(t)} - (A_{0\sim n-1}E(t))_j\right] - \frac{h_{n+1,j}(e_1)}{\Gamma(t)},
\end{aligned}
\tag{2.17}
$$

where

$$
G(X,\xi) = \mathrm{diag}(G_1, G_2, \cdots, G_m), \tag{2.18}
$$
$$
K(X,\xi) = \mathrm{diag}(K_1, K_2, \cdots, K_m), \tag{2.19}
$$

and the concrete form of $E = [E_1^\mathsf{T}, E_2^\mathsf{T}, \cdots, E_n^\mathsf{T}]^\mathsf{T}$ is

$$
\begin{cases}
E_i = \hat{x}_{i+1} + \Gamma^{n-i}(t)h_i(e_1) - x_{R(i+1)}, i \in [n-1], \\
E_n = \hat{\varpi} + h_n(e_1) + \bar{G}u - x_{R(n+1)}.
\end{cases}
\tag{2.20}
$$

Define the tracking error

$$
\varepsilon_i = x_i - x_{Ri}, i \in [n], \tag{2.21}
$$

and the derivatives of $\varepsilon_i, i \in [n]$ are

$$
\begin{cases}
\dot{\varepsilon}_i = x_{i+1} - x_{R(i+1)} = \varepsilon_{i+1}, i \in [n-1], \\
\dot{\varepsilon}_n = \varpi + \bar{G}u - x_{R(n+1)} \\
\quad = e_{n+1} - A_{0\sim n-1}(\hat{X} - X_R) \\
\quad = e_{n+1} - A_{0\sim n-1}(X - X_R) + A_{0\sim n-1}(X - \hat{X}).
\end{cases}
\tag{2.22}
$$

Let $\mathscr{E}_O = [e_1^\mathsf{T}, e_2^\mathsf{T}, \cdots, e_{n+1}^\mathsf{T}]^\mathsf{T}$ and $\mathscr{E}_T = [\varepsilon_1^\mathsf{T}, \varepsilon_2^\mathsf{T}, \cdots, \varepsilon_n^\mathsf{T}]^\mathsf{T}$, it holds that $\|X - \hat{X}\| \leq \|\mathscr{E}_O\|$. Based on Assumptions 2.1–2.2, the boundedness of faults, the convergence of TD and the characteristics of nonlinear functions h_i, there exist positive scalars M_0, M_1, M_2, M_c such that

$$\|\dot{\varpi}\| \leq M_0 + M_1\|\mathscr{E}_O\| + M_2\|\mathscr{E}_T\| + \frac{\mathscr{H}_{n+1}}{\epsilon}M\|\mathscr{E}_O\|, \tag{2.23}$$

where

$$M = \max_{j\in[m]} \left\| \frac{G_j - g}{g} \right\| \leq M_c < \infty. \tag{2.24}$$

The overall error dynamics is

$$\begin{cases} \dot{\varepsilon}_i = \varepsilon_{i+1}, i \in [n-1], \\ \dot{\varepsilon}_n = e_{n+1} - A_{0\sim n-1}\mathscr{E}_T + A_{0\sim n-1}(X - \hat{X}), \\ \dot{e}_i = \dfrac{e_{i+1}}{\Gamma(t)} - \dfrac{h_i(e_1)}{\Gamma(t)} - \dfrac{(n+1-i)\dot{\Gamma}(t)e_i}{\Gamma(t)}, i \in [n], \\ \dot{e}_{n+1} = \dot{\varpi} - \dfrac{1}{\Gamma(t)}h_{n+1}(e_1). \end{cases} \tag{2.25}$$

Consider the following Lyapunov function

$$\begin{aligned} V &= V_O(\mathscr{E}_O) + V_T(\mathscr{E}_T) \\ &= V_O(\mathscr{E}_O) + \mathscr{E}_T^\mathsf{T} P \mathscr{E}_T, \end{aligned} \tag{2.26}$$

where V_O comes from Assumption 2.3, and P comes from the parameterized matrix in Chap. 1 satisfying

$$\Phi_A^\mathsf{T} P + P\Phi_A = -\kappa_1 I_{mn}, \tag{2.27}$$

$$\Phi_A = \begin{bmatrix} 0 & I_m & 0 & \cdots & 0 \\ 0 & 0 & I_m & \cdots & 0 \\ \vdots & & & \ddots & \vdots \\ 0 & 0 & 0 & \cdots & I_m \\ -A_0 & -A_1 & -A_2 & \cdots & -A_{n-1} \end{bmatrix}. \tag{2.28}$$

κ_1 is the positive scalar to be determined.

Naturally, it turns out that

$$\lambda_{\min}(P)\|\mathscr{E}_T\|^2 \leq V_T(\mathscr{E}_T) \leq \lambda_{\max}(P)\|\mathscr{E}_T\|^2, \tag{2.29}$$

$$\sum_{i=1}^{n-1}[\frac{\partial V_T}{\partial \varepsilon_i}(\mathscr{E}_T)]^{\mathsf{T}}\varepsilon_{i+1} - \sum_{i=1}^{n}[\frac{\partial V_T}{\partial \varepsilon_n}(\mathscr{E}_T)]^{\mathsf{T}}A_{i-1}\varepsilon_i \tag{2.30}$$

$$= \mathscr{E}_T^{\mathsf{T}}\left(\Phi_A^{\mathsf{T}}P + P\Phi_A\right)\mathscr{E}_T = -\kappa_1\|\mathscr{E}_T\|^2,$$

$$\left\|\frac{\partial V_T}{\partial \varepsilon_n}(\mathscr{E}_T)\right\| \le 2\lambda_{\max}(P)\|\mathscr{E}_T\|. \tag{2.31}$$

Combining Assumption 2.3, $\|-\dot{\Gamma}(t)/\Gamma(t)\| \le \mathfrak{a}$, $\Gamma(t) \ge \epsilon_0$, (2.23) and (2.29), the derivative of V is

$$\frac{dV}{dt} = \frac{1}{\Gamma(t)}\left\{\sum_{i=1}^{n}[\frac{\partial V_O}{\partial e_i}(\mathscr{E}_O)]^{\mathsf{T}}(e_{i+1} - h_i(e_1))\right.$$

$$\left. +[\frac{\partial V_O}{\partial e_{n+1}}(\mathscr{E}_O)]^{\mathsf{T}}(\Gamma(t)\dot{\varpi} - h_{n+1}(e_1))\right\}$$

$$-\frac{\dot{\Gamma}(t)}{\Gamma(t)}\sum_{i=1}^{n}(n+1-i)[\frac{\partial V_O}{\partial e_i}(\mathscr{E}_O)]^{\mathsf{T}}e_i + \sum_{i=1}^{n-1}\frac{\partial V_T}{\partial \varepsilon_i}(\mathscr{E}_T)]^{\mathsf{T}}\varepsilon_{i+1}$$

$$+[\frac{\partial V_T}{\partial \varepsilon_n}(\mathscr{E}_T)]^{\mathsf{T}}\left[e_{n+1} - A_{0\sim n-1}\mathscr{E}_T + A_{0\sim n-1}(X - \hat{X})\right]$$

$$< \frac{1}{\Gamma(t)}\left[-W_O(\mathscr{E}_O) + \Gamma(t)\left(M_1\|\mathscr{E}_O\| + M_2\|\mathscr{E}_T\| + M_0\right.\right.$$

$$\left.\left. +\frac{\mathscr{H}_{n+1}}{\epsilon}M\|\mathscr{E}_O\|)\beta\|\mathscr{E}_O\|\right]$$

$$+ n\mathfrak{a}\gamma W_O(\mathscr{E}_O) - \kappa_1\|\mathscr{E}_T\|^2 + 2\lambda_{\max}(P)(\|A_{0\sim n-1}\| + 1)\|\mathscr{E}_O\|\|\mathscr{E}_T\|$$

$$\le -\left(\frac{a_1}{\epsilon} - \beta M_1 - n\mathfrak{a}\gamma\alpha_4\right)\|\mathscr{E}_O\|^2$$

$$+ \beta M_0\|\mathscr{E}_O\| - \kappa_1\|\mathscr{E}_T\|^2 + a_2\|\mathscr{E}_O\|\|\mathscr{E}_T\|$$

$$\le -\left(\frac{a_1}{2\epsilon} - \beta M_1 - n\mathfrak{a}\gamma\alpha_4\right)\|\mathscr{E}_O\|^2 + \beta M_0\|\mathscr{E}_O\| - \left(\kappa_1 - \frac{\epsilon a_2^2}{2a_1}\right)\|\mathscr{E}_T\|^2, \tag{2.32}$$

where $a_1 = \alpha_3 - \beta\mathscr{H}_{n+1}M_c > 0$, $a_2 = \beta M_2 + 2\lambda_{\max}(P)(\|A_{0\sim n-1}\| + 1)$.
   Define two constants

$$\begin{cases}\Theta = \max\left\{2, \dfrac{4\beta M_0}{\kappa_1}\right\}, \\[2mm] \epsilon_0 = \min\left\{1, \dfrac{a_1}{2\beta(M_0 + M_1) + 2n\mathfrak{a}\gamma\alpha_4}, \dfrac{a_1\kappa_1}{a_2^2}\right\}.\end{cases} \tag{2.33}$$

If $\epsilon \in (0, \epsilon_0)$, the overall error dynamics is ultimately uniformly bounded, i.e., $\exists t_\epsilon > 0, \|[\mathscr{E}_O^{\mathsf{T}}, \mathscr{E}_T^{\mathsf{T}}]^{\mathsf{T}}\| \leq \Theta, \forall t \geq t_\epsilon$. Substituting the bounded error into (2.23), there exists a positive scalar M_Θ satisfying $\dot{\varpi} \leq M_\Theta + \mathscr{H}_{n+1} M_c \|\mathscr{E}_O\|/\epsilon, \forall t > t_\epsilon$. Then, it can be obtained that

$$
\begin{aligned}
\frac{\mathrm{d}V_O}{\mathrm{d}t} &= \frac{1}{\Gamma(t)} \left\{ \sum_{i=1}^{n} [\frac{\partial V_O}{\partial e_i}(\mathscr{E}_O)]^{\mathsf{T}} (e_{i+1} - h_i(e_1)) \right. \\
&\quad \left. + [\frac{\partial V_O}{\partial e_{n+1}}(\mathscr{E}_O)]^{\mathsf{T}} (\Gamma(t)\dot{\varpi} - h_{n+1}(e_1)) \right\} \\
&\quad - \frac{\dot{\Gamma}(t)}{\Gamma(t)} \sum_{i=1}^{n} (n+1-i)[\frac{\partial V_O}{\partial e_i}(\mathscr{E}_O)]^{\mathsf{T}} e_i \\
&\leq -\frac{a_1^*}{\epsilon} \|\mathscr{E}_O\|^2 + M_\Theta \beta \|\mathscr{E}_O\| \\
&\leq -\frac{a_1^*}{\epsilon \alpha_2} V_O(\mathscr{E}_O) + M_\Theta \beta \sqrt{\frac{V_O(\mathscr{E}_O)}{\alpha_1}},
\end{aligned}
\tag{2.34}
$$

where $a_1^* = a_1 - \epsilon n a \gamma \alpha_4 > 0$. Moreover, the above equation can be transformed into

$$
\frac{\mathrm{d}\sqrt{V_O(\mathscr{E}_O)}}{\mathrm{d}t} \leq -\frac{a_1^*}{2\epsilon \alpha_2} \sqrt{V_O(\mathscr{E}_O)} + \frac{M_\Theta \beta}{2\sqrt{\alpha_1}}.
\tag{2.35}
$$

Thus, for $\forall t > t_\epsilon$, it holds that

$$
\sqrt{V_O(\mathscr{E}_O(t))} \leq \sqrt{V_O(\mathscr{E}_O(t_\epsilon))} \exp(-\frac{a_1^*}{2\epsilon \alpha_2}(t - t_\epsilon)) + \frac{M_\Theta \beta}{2\sqrt{\alpha_1}} \int_{t_\epsilon}^{t} e^{-\frac{a_1^*}{2\epsilon \alpha_2}(t-\tau)} \, \mathrm{d}\tau.
\tag{2.36}
$$

The above result implies that there exist $t_{1\epsilon} > t_\epsilon$ and a positive scalar Θ_1 such that for $\forall t \geq t_{1\epsilon}, i \in [n]$, the following equations hold.

$$
\begin{cases}
\|x_i - \hat{x}_i\| = \epsilon^{n+1-i} \|e_i\| \leq \epsilon^{n+1-i} \|\mathscr{E}_O\| \leq \epsilon^{n+1-i} \sqrt{\frac{V_O(\mathscr{E}_O)}{\alpha_1}} \leq \epsilon^{n+2-i} \Theta_1, \\
\|\varpi - \hat{\varpi}\| = \|e_{n+1}\| \leq \epsilon \Theta_1.
\end{cases}
\tag{2.37}
$$

Based on (2.37), calculating the derivative of V_T yields

$$
\frac{\mathrm{d}V_T}{\mathrm{d}t} = \sum_{i=1}^{n-1} [\frac{\partial V_T}{\partial \varepsilon_i}(\mathscr{E}_T)]^{\mathsf{T}} \varepsilon_{i+1} + [\frac{\partial V_T}{\partial \varepsilon_n}(\mathscr{E}_T)]^{\mathsf{T}}
$$

$$\times \left[e_{n+1} - A_{0\sim n-1}\mathscr{E}_T + A_{0\sim n-1}(X - \hat{X}) \right]$$

$$\leq -\kappa_1 \|\mathscr{E}_T\|^2 + (a_2 - \beta M_2)\|\mathscr{E}_O\|\|\mathscr{E}_T\|$$

$$\leq -\frac{\kappa_1}{\lambda_{\max}(P)} V_T(\mathscr{E}_T) + \epsilon M_{\Theta_1}\sqrt{V_T(\mathscr{E}_T)}, \tag{2.38}$$

where M_{Θ_1} is a positive scalar depending on R. Then, it can be obtained that

$$\frac{d\sqrt{V_T(\mathscr{E}_T)}}{dt} \leq -\frac{\kappa_1}{2\lambda_{\max}(P)}\sqrt{V_T(\mathscr{E}_T)} + \frac{\epsilon M_{\Theta_1}}{2}. \tag{2.39}$$

For $\forall t > t_{1\epsilon}$, it holds that

$$\sqrt{V_T(\mathscr{E}_T(t))} \leq \sqrt{V_T(\mathscr{E}_T(t_{1\epsilon}))}e^{-\frac{\kappa_1}{2\lambda_{\max}(P)}(t-t_{1\epsilon})} + \frac{\epsilon M_{\Theta 1}}{2}\int_{t_{1\epsilon}}^t e^{-\frac{\kappa_1}{2\lambda_{\max}(P)}(t-\tau)}\,d\tau. \tag{2.40}$$

It means

$$\|\mathscr{E}_T(t)\| \leq \frac{1}{\sqrt{\lambda_{\min}(P)}}\left(\sqrt{V_T(\mathscr{E}_T(t_{1\epsilon}))}e^{-\frac{\kappa_1}{2\lambda_{\max}(P)}(t-t_{1\epsilon})}\right.$$
$$\left. + \frac{\epsilon M_{\Theta 1}}{2}\int_{t_{1\epsilon}}^t e^{-\frac{\kappa_1}{2\lambda_{\max}(P)}(t-\tau)}\,d\tau\right). \tag{2.41}$$

Therefore, there exists $t_{2\epsilon} > t_{1\epsilon}$ such that

$$\|x_i - x_{Ri}\| \leq \epsilon\Theta_2, t \geq t_{2\epsilon}, i \in [n]. \tag{2.42}$$

According to the convergence of TD given by Lemma 2.1, there exist positive scalars $t_{R\epsilon} > 0$ and Θ_3 such that

$$\|y - y_d\| \leq \epsilon\Theta_3, t \geq t_{R\epsilon}. \tag{2.43}$$

The proof is completed.

Corollary 2.1 *If the functions $h_i(r)$ are chosen as $h_i(r) = \mathscr{H}_i r, i \in [n+1]$, the observer and controller integration can be designed as*

$$\begin{cases} \dot{\hat{x}}_i = \hat{x}_{i+1} + \frac{\mathscr{H}_i}{\Gamma^i(t)}(y - \hat{x}_1), i \in [n-1], \\ \dot{\hat{x}}_n = \hat{\varpi} + \frac{\mathscr{H}_n}{\Gamma^n(t)}(y - \hat{x}_1) + \bar{G}u, \\ \dot{\hat{\varpi}} = \frac{\mathscr{H}_{n+1}}{\Gamma^{n+1}(t)}(y - \hat{x}_1), \\ u = -\bar{G}^{-1}\left[\hat{\varpi} - x_{R(n+1)} + A_{0\sim n-1}(\hat{X} - X_R)\right]. \end{cases} \tag{2.44}$$

When Assumptions 2.1 and 2.2 hold, there exist positive scalars $t_{1\epsilon}$, $t_{2\epsilon}$, $t_{R\epsilon}$, Θ_1, Θ_2, Θ_3 such that

1. *The observation error and tracking error are ultimately uniformly bounded.*
2. *For any $\epsilon \in (0, \epsilon_0)$, it holds that*

$$\begin{cases} \|x_i - \hat{x}_i\| \leq \epsilon^{n+2-i}\Theta_1, t \geq t_{1\epsilon}, i \in [n], \\ \|\varpi - \hat{\varpi}\| \leq \epsilon\Theta_1, t \geq t_{1\epsilon} \\ \|x_i - x_{Ri}\| \leq \epsilon\Theta_2, t \geq t_{2\epsilon}, i \in [n], \\ \|y - y_d\| \leq \epsilon\Theta_3, t \geq t_{R\epsilon}. \end{cases} \tag{2.45}$$

Proof Compared with Theorm 2.1, Corollary 2.1 does not demand Assumption 2.3. The reason is that the integration framework (2.44) is consistent with Assumption 2.3. Define a Hurwitz matrix

$$\Psi = \begin{bmatrix} -\mathscr{H}_1 I_m & I_m & 0 & \cdots & 0 \\ -\mathscr{H}_2 I_m & 0 & I_m & \cdots & 0 \\ \vdots & & & \ddots & \\ -\mathscr{H}_n I_m & 0 & 0 & \cdots & I_m \\ -\mathscr{H}_{n+1} I_m & 0 & 0 & \cdots & 0 \end{bmatrix}, \tag{2.46}$$

$\mathscr{P} \in \mathbb{R}^{m(n+1) \times m(n+1)}$ is a positive definite solution to the following Lyapunov equation

$$\mathscr{P}\Psi + \Psi^{\mathsf{T}}\mathscr{P} = -I_{m(n+1)}, \tag{2.47}$$

and $\mathscr{P}$ satisfies

$$M\mathscr{H}_{n+1} < 1/[2\lambda_{\max}(\mathscr{P})]. \tag{2.48}$$

Consider the Lyapunov function

$$V_O(\mathfrak{z}) = \mathfrak{z}^{\mathsf{T}}\mathscr{P}\mathfrak{z}, W_O(\mathfrak{z}) = \mathfrak{z}^{\mathsf{T}}\mathfrak{z}, \forall \mathfrak{z} = [\mathfrak{z}_1^{\mathsf{T}}, \mathfrak{z}_2^{\mathsf{T}}, \cdots, \mathfrak{z}_{n+1}^{\mathsf{T}}]^{\mathsf{T}}, \tag{2.49}$$

and there are some conclusions

$$\begin{cases} \lambda_{\min}(\mathscr{P})\|\mathfrak{z}\|^2 \leq V_O(\mathfrak{z}) \leq \lambda_{\max}(\mathscr{P})\|\mathfrak{z}\|^2, \\ \sum_{i=1}^{n}[\frac{\partial V_O}{\partial \mathfrak{z}_i}(\mathfrak{z})]^{\mathsf{T}}(\mathfrak{z}_{i+1} - \mathscr{H}_i\mathfrak{z}_1) - [\frac{\partial V_O}{\partial \mathfrak{z}_{n+1}}(\mathfrak{z})]^{\mathsf{T}}\mathscr{H}_{n+1}\mathfrak{z}_1 = -W_O(\mathfrak{z}), \\ \left\|\frac{\partial V_O}{\partial \mathfrak{z}_{n+1}}(\mathfrak{z})\right\| \leq 2\lambda_{\max}(\mathscr{P})\|\mathfrak{z}\|, \\ \sum_{i=1}^{n}[\frac{\partial V_O}{\partial \mathfrak{z}_i}(\mathfrak{z})]^{\mathsf{T}}\mathfrak{z}_i \leq 2\lambda_{\max}(\mathscr{P})W_O(\mathfrak{z}). \end{cases} \tag{2.50}$$

Obviously, the corresponding conditions in Assumption 2.3 hold. The remaining proofs are similar to Theorem 2.1.

Remark 2.3 The two Hurwitz matrices Φ_A, Ψ indicate that the closed-loop FAS model meets the generalized duality principle. The main difference is that the dimensions of the two matrices do not agree, with Ψ one extra dimension corresponding to the total system nonlinearity. ϖ includes the component faults, and $\hat{\varpi}$ adaptively estimates the total uncertainty including the faults, so such an integration framework is a passive fault-tolerant control (PFTC) strategy.

2.3.2 Dynamical Data Modeling

The integration design (2.11) can give the following dynamics

$$
\begin{cases}
\dot{\hat{x}}_n = \hat{\varpi} + h_n(\vartheta) + \bar{G}u, \\
\dot{\hat{\varpi}} = \dfrac{1}{\Gamma(t)} h_{n+1}(\vartheta), \\
u = -\bar{G}^{-1}\left[\hat{\varpi} - x_{R(n+1)} + A_{0\sim n-1}(\hat{X} - X_R)\right].
\end{cases}
\tag{2.51}
$$

The corresponding closed-loop system is

$$
\dot{\hat{x}}_n = x_{R(n+1)} - A_{0\sim n-1}(\hat{X} - X_R) + h_n(\vartheta).
\tag{2.52}
$$

Let $S = [s_1^\mathsf{T}, s_2^\mathsf{T}, \cdots, s_n^\mathsf{T}]^\mathsf{T}$, $s_i = \hat{x}_i - x_{Ri}$, $i \in [n]$, $s_{n+1} = \hat{\varpi} - x_{R(n+1)}$, a data model of the closed-loop FAS model is

$$
\begin{cases}
\dot{S} = \Phi_A S + \mathfrak{D}_1, \mathfrak{D}_1 = [\Gamma^{n-1}(t)h_1(\vartheta)^\mathsf{T}, \Gamma^{n-2}(t)h_2(\vartheta)^\mathsf{T}, \cdots, h_n(\vartheta)^\mathsf{T}]^\mathsf{T}, \\
u = -\bar{G}^{-1}(s_{n+1} + A_{0\sim n-1}S), \\
\dot{s}_{n+1} = \dfrac{1}{\Gamma(t)} h_{n+1}(\vartheta) - R^n \varphi(x_{R1} - y_d, \frac{x_{R2}}{R}, \cdots, \frac{x_{R(n+1)}}{R^n}),
\end{cases}
\tag{2.53}
$$

where $\vartheta = (y - s_1 - x_{R1})/\Gamma^n(t)$, and s_{n+1} is the total uncertainty. Obviously, the output y determine the whole dynamic data model of (2.6).

Theorem 2.2 *The new form (2.53) is a closed-loop dynamic data modeling of the tracking error dynamics, i.e., there exist positive scalars $t_{s\epsilon}$, Θ_4 such that*

$$
\|x^{(i-1)} - y_d^{(i-1)} - s_i\| \le \epsilon\Theta_4, t \ge t_{s\epsilon}, i \in [n].
\tag{2.54}
$$

Proof The tracking error dynamics of (2.6) is

$$x^{(n)} - y_d^{(n)} = F(x^{(0\sim n-1)}, \xi) + G(x^{(0\sim n-1)}, \xi)u + K(x^{(0\sim n-1)}, \xi)d + \delta - y_d^{(n)}. \tag{2.55}$$

Model-driven robust control methods require $x^{(0\sim n-1)}$, ξ, $F(\cdot)$, $G(\cdot)$ and the upper bound information for disturbances and faults. On the condition that $\|D(t)\| \le D_m$ where D_m is a positive scalar, $P = [P_1, P_2, \cdots, P_n]$, and the auxiliary signal is

$$u_a = 1/(4\epsilon)(D_m^2 + \delta_m^2) P_n^{\mathsf{T}}(x^{(0\sim n-1)} - y_d^{(0\sim n-1)}). \tag{2.56}$$

The corresponding model-driven robust control law is [25]

$$u = -G^{-1}(x^{(0\sim n-1)})[F(x^{(0\sim n-1)}) + u_a + A_{0\sim n-1}(x^{(0\sim n-1)} - y_d^{(0\sim n-1)}) - y_d^{(n)}]. \tag{2.57}$$

Define $\eta = x - y_d$, and then the closed-loop FAS tracking error model (2.6) is

$$\dot{\eta}^{(0\sim n-1)} = \Phi_A \eta^{(0\sim n-1)} + \mathfrak{D}_2, \mathfrak{D}_2 = [0_{m(n-1)}^{\mathsf{T}}, (D(t) + \delta - u_a)^{\mathsf{T}}]^{\mathsf{T}}. \tag{2.58}$$

The data modeling error is defined as $\zeta = \eta^{(0\sim n-1)} - S$, and it turns out that

$$\dot{\zeta} = \Phi_A \zeta + \mathfrak{D}_2 - \mathfrak{D}_1. \tag{2.59}$$

When $t \ge -\ln \epsilon/\mathfrak{a}$, the derivative of a Lyapunov function $V_\zeta = \frac{1}{2}\zeta^{\mathsf{T}} P \zeta$ is

$$\begin{aligned}
\frac{dV_\zeta}{dt} &= \frac{1}{2}(\Phi_A \zeta + \mathfrak{D}_2 - \mathfrak{D}_1)^{\mathsf{T}} P \zeta + \frac{1}{2}\zeta^{\mathsf{T}} P (\Phi_A \zeta + \mathfrak{D}_2 - \mathfrak{D}_1) \\
&= \frac{1}{2}\zeta^{\mathsf{T}}(\Phi_A^{\mathsf{T}} P + P \Phi_A)\zeta + \zeta^{\mathsf{T}} P(\mathfrak{D}_2 - \mathfrak{D}_1) \\
&< -\frac{\mu}{2}V_\zeta + \zeta^{\mathsf{T}} P(\mathfrak{D}_2 - \mathfrak{D}_1). \tag{2.60}
\end{aligned}$$

It can be calculated that

$$\begin{aligned}
\zeta^{\mathsf{T}} P(\mathfrak{D}_2 - \mathfrak{D}_1) &= \zeta^{\mathsf{T}} P_n(D + \delta - v) - \zeta(t)^{\mathsf{T}} P \mathfrak{D}_1 \\
&= \zeta^{\mathsf{T}} P_n(D + \delta - \frac{D_m^2 + \delta_m^2}{4\epsilon} P_n^{\mathsf{T}} \eta^{(0\sim n-1)}) - \zeta^{\mathsf{T}} P \mathfrak{D}_1 \\
&= -\frac{D_m^2 + \delta_m^2}{4\epsilon}\zeta^{\mathsf{T}} P_n P_n^{\mathsf{T}} \zeta - \frac{D_m^2 + \delta_m^2}{4\epsilon}\zeta^{\mathsf{T}} P_n P_n^{\mathsf{T}} S \\
&\quad + \zeta^{\mathsf{T}} P_n(D + \delta) - \zeta^{\mathsf{T}} P \mathfrak{D}_1 \\
&= -\frac{D_m^2 + \delta_m^2}{4\epsilon}\|P_n^{\mathsf{T}} \zeta\|^2 - \zeta^{\mathsf{T}} P_n \mathfrak{D}_3, \tag{2.61}
\end{aligned}$$

where

$$\mathfrak{D}_3 = D + \delta - \mathfrak{D}_1 - \frac{D_m^2 + \delta_m^2}{4\epsilon} P_n^\mathsf{T} S. \tag{2.62}$$

Based on Assumption 2.1 and Theorem 2.1, it turns out that

$$\|\mathfrak{D}_3\| \leq \|D\| + \|\delta\| + \|\mathfrak{D}_1\| + \left\| \frac{D_m^2 + \delta_m^2}{4\epsilon} P_n^\mathsf{T} S \right\|$$

$$\leq D_m + \delta_m + \epsilon \Theta_{\mathfrak{D}} + \frac{D_m^2 + \delta_m^2}{4} \Theta_S \|P_n\|$$

$$= \epsilon \Theta_{\mathfrak{D}} + \Theta_{\mathfrak{P}}, \quad t > t_{2\epsilon}, \tag{2.63}$$

where Θ_S, $\Theta_{\mathfrak{D}}$, $\Theta_{\mathfrak{P}}$ are positive scalars. Then, it follows that

$$\zeta^\mathsf{T} P(\mathfrak{D}_2 - \mathfrak{D}_1) \leq -\frac{D_m^2}{4\epsilon} \|P_n^\mathsf{T} \zeta\|^2 + (\epsilon \Theta_{\mathfrak{D}} + \Theta_{\mathfrak{P}}) \|P_n^\mathsf{T} \zeta\|$$

$$\leq \frac{\epsilon(\epsilon \Theta_{\mathfrak{D}} + \Theta_{\mathfrak{P}})^2}{D_m^2}, \quad t > t_{2\epsilon}. \tag{2.64}$$

Combining (2.60) and (2.64), it can be obtained that

$$\frac{dV_\zeta}{dt} < -\frac{\mu}{2} V_\zeta + \frac{\epsilon(\epsilon \Theta_{\mathfrak{D}} + \Theta_{\mathfrak{P}})^2}{D_m^2}, \tag{2.65}$$

i.e., $V_\zeta \to \epsilon(\epsilon \Theta_{\mathfrak{D}} + \Theta_{\mathfrak{P}})^2 / D_m^2$, $t \to \infty$. Therefore, the dynamic data modeling error is bounded, and there are positive scalars $t_{s\epsilon}$, Θ_4 such that

$$\|x^{(i-1)} - y_d^{(i-1)} - s_i\| \leq \epsilon \Theta_4, \quad t \geq t_{s\epsilon}, \quad i \in [n]. \tag{2.66}$$

The proof is completed.

Corollary 2.2 *If the functions $h_i(r)$ are chosen as $h_i(r) = \mathcal{H}_i r$, $i \in [n+1]$, the new form (2.67) is a closed-loop dynamic data modeling of the tracking error dynamics.*

$$\dot{S} = \Phi_A S + \mathfrak{D}_1, \quad \mathfrak{D}_1 = \begin{bmatrix} \frac{\mathcal{H}_1}{\Gamma(t)}(y - s_1 - x_{R1}) \\ \frac{\mathcal{H}_2}{\Gamma^2(t)}(y - s_1 - x_{R1}) \\ \vdots \\ \frac{\mathcal{H}_n}{\Gamma^n(t)}(y - s_1 - x_{R1}) \end{bmatrix}. \tag{2.67}$$

There exist positive scalars $t_{s\epsilon}$, Θ_4 such that

$$\|x^{(i-1)} - x_d^{(i-1)} - s_i\| \le \epsilon \Theta_4, t \ge t_{s\epsilon}, i \in [n]. \tag{2.68}$$

Proof The proof is similar to Theorem 2.2.

Remark 2.4 Different from the stability proof of Theorems 2.1 and 2.2 indicates that the control performance under the data modeling mechanism is basically the same as that of the model-driven controller. The physical meaning of Theorem 2.2 is that the closed-loop data model (2.53) can effectively approximate the FAS model with uncertain dynamics.

2.4   Active Fault-Tolerant Control Design

Based on Theorem 2.2, the diagnosis and control problem of uncertain high-order FAS (2.6) is transformed into the diagnosis and control problem of dynamic data models (2.51) and (2.53). The closed-loop data model is essentially a linear system, and the mature linear system theory can be utilized. Naturally, the original study of complex systems is reasonably transformed into a simple linear system study.

The controller (2.11) is actually a PFTC policy because $\hat{\varpi}$ continuously estimates the total uncertainty including component faults. If the output data y comes from a faulty system, then the corresponding dynamic data model is a PFTC model. If the output data $\tilde{y}$ comes from a fault-free system, then the corresponding dynamic data model can approximate the fault-free system model. However, the control performance of PFTC is weaker than that of AFTC. Especially in the case of more serious faults, the general PFTC may not be able to compensate faults in time. Therefore, this section considers the AFTC design of the FAS model: the first step is to solve the fault diagnosis design of the closed-loop FAS model, and the second step is to solve the controller reconfiguration rules.

Based on Theorem 2.1, the integration (2.11) can give

$$\begin{cases} \tilde{\varpi} = \varpi - \delta - \varXi(t), t \ge t_{1\epsilon}, \\ \|\varpi - \hat{\varpi}\| \le \epsilon \Theta_1, t \ge t_{1\epsilon}, \\ \|\tilde{\varpi} - \hat{\tilde{\varpi}}\| \le \epsilon \Theta_1, t \ge t_{1\epsilon}, \end{cases} \tag{2.69}$$

where

$$\hat{\tilde{\varpi}} = \frac{1}{\Gamma(t)} h_{n+1}\left(\frac{\tilde{y} - \hat{\tilde{x}}_1}{\Gamma^n(t)}\right). \tag{2.70}$$

$\tilde{\varpi}, \varpi$ are the total uncertainty of the fault-free FAS and the faulty FAS, respectively. $\hat{\tilde{x}}_i$ is the state observation determined by the output $\tilde{y}$ from the fault-free FAS. $\varXi(t)$ is the remaining term caused by disturbances between the two data models.

Remark 2.5 $\varXi(t)$ meets $\sup_{t\ge0} \|\varXi(t)\| \le \varXi_m$ where $\varXi_m$ is a positive scalar. For the same system with the same tracking control task, the remaining term $\varXi(t)$

obtained from multiple measurements is usually small. In particular, the remaining term $\varXi(t)$ is rather small when the external disturbance has a dynamic structure.

2.4.1 Fault Diagnosis Framework

If $\|\hat{\tilde{\omega}} - \hat{\omega}\| > \mathfrak{e}_{th}, t \geq t_{th}$, there are component faults in the nonlinear FAS (2.6) model. $\mathfrak{e}_{th}, t_{th}$ are the detection threshold and detection time, respectively. In order to show the fault detection logic more clearly, the dynamic data models of the fault-free FAS and the faulty FAS are as follows:

$$\begin{cases} \dot{\tilde{S}} = \varPhi_A \tilde{S} + \tilde{\mathfrak{D}}_1, \tilde{\mathfrak{D}}_1 = [\varGamma^{n-1}(t)h_1(\tilde{\vartheta})^\mathsf{T}, \varGamma^{n-2}(t)h_2(\tilde{\vartheta})^\mathsf{T}, \cdots, h_n(\tilde{\vartheta})^\mathsf{T}]^\mathsf{T}, \\ \tilde{u} = -\bar{G}^{-1}(\tilde{s}_{n+1} + A_{0 \sim n-1}\tilde{S}), \\ \dot{\tilde{s}}_{n+1} = \varGamma^{-1}(t)h_{n+1}(\tilde{\vartheta}) - R^n\varphi(x_{R1} - y_d, \frac{x_{R2}}{R}, \cdots, \frac{x_{R(n+1)}}{R^n}), \\ \tilde{\vartheta} = (\tilde{y} - \tilde{s}_1 - x_{R1})/\varGamma^n(t), \end{cases} \tag{2.71}$$

$$\begin{cases} \dot{S} = \varPhi_A S + \mathfrak{D}_1, \mathfrak{D}_1 = [\varGamma^{n-1}(t)h_1(\vartheta)^\mathsf{T}, \varGamma^{n-2}(t)h_2(\vartheta)^\mathsf{T}, \cdots, h_n(\vartheta)^\mathsf{T}]^\mathsf{T}, \\ u = -\bar{G}^{-1}(s_{n+1} + A_{0 \sim n-1}S), \\ \dot{s}_{n+1} = \varGamma^{-1}(t)h_{n+1}(\vartheta) - R^n\varphi(x_{R1} - y_d, \frac{x_{R2}}{R}, \cdots, \frac{x_{R(n+1)}}{R^n}), \\ \vartheta = (y - s_1 - x_{R1})/\varGamma^n(t). \end{cases} \tag{2.72}$$

The two models are derived from the output data $\tilde{y}$ of the pre-stored fault-free system and the output data y of the real-time system, which may encounter component faults.

Theorem 2.3 *For the faulty FAS model* (2.6), *the following conclusions hold.*

1. The fault detectability condition is

$$\begin{cases} \|\delta\| > \varXi_m, t \in [t_1, t_2], t_2 > t_1 > 0, \\ \|s_{n+1} - \tilde{s}_{n+1}\| > \mathfrak{e}_{th}, t \geq t_{th} > t_2, \end{cases} \tag{2.73}$$

where $\tilde{s}_{n+1}, s_{n+1}$ *are derived from the fault-free system data model and the real-time system data model, respectively. The first inequality of* (2.73) *indicates that the fault amplitude can only be detected if it exceeds the maximum of the remaining disturbance term.*

2. The fault estimation $\hat{\delta}$ *can be calculated as*

$$\hat{\delta} = s_{n+1} - \tilde{s}_{n+1}. \tag{2.74}$$

The fault estimation error is ultimately uniformly bounded, i.e.,

$$\|\delta - \hat{\delta}\| \leq 2\epsilon\Theta_1 + \Xi_m, t \geq t_{1\epsilon}. \tag{2.75}$$

Proof Define the residual signal of the faulty system (2.6)

$$\mathfrak{e} = s_{n+1} - \tilde{s}_{n+1}. \tag{2.76}$$

If the corresponding data model parameters are the same, the derivative of the residual signal is

$$\dot{\mathfrak{e}} = \dot{s}_{n+1} - \dot{\tilde{s}}_{n+1} = \frac{1}{\epsilon}(h_{n+1}(\vartheta) - h_{n+1}(\tilde{\vartheta})) \begin{cases} \approx 0, t_s < t < t_f, \\ \neq 0, t_f \leq t < t_{pf}, \\ \approx 0, t \geq t_{pf}, \end{cases} \tag{2.77}$$

where t_s, t_f, t_{pf} are the stabilization time of the fault-free system, fault occurrence time, and PFTC completion time, respectively. Before the fault occurrence and after the PFTC completion, ϑ and $\tilde{\vartheta}$ are very close. In the PFTC process, ϑ and $\tilde{\vartheta}$ are different. $\|\delta\| > \Xi_m$ is a guarantee that the fault can be detected, and serious faults meet this condition.

Therefore, the fault detection condition is $\|s_{n+1} - \tilde{s}_{n+1}\| > \mathfrak{e}_{th}, t \geq t_{th}$. There are two ways to set the detection threshold $\mathfrak{e}_{th}$: (1) When the disturbance information is accurate, the remaining disturbance item $\Xi(t)$ is determined according to mechanism models or engineering experience, and the detection threshold should exceed the remaining item $\Xi(t)$. (2) When the disturbance information is not accurate, Monte Carlo experiments can give the average value of the remaining $\Xi(t)$, and the detection threshold should exceed the average value.

Based on Theorem 2.1, it turns out that

$$\begin{cases} \|\varpi - s_{n+1} - x_{R(n+1)}\| \leq \epsilon\Theta_1, t \geq t_{1\epsilon}, \\ \|\tilde{\varpi} - \tilde{s}_{n+1} - x_{R(n+1)}\| \leq \epsilon\Theta_1, t \geq t_{1\epsilon}. \end{cases} \tag{2.78}$$

The fault estimation error is defined as $\tilde{\delta} = \delta - \hat{\delta}$, and it follows that

$$\tilde{\delta} = \varpi - \tilde{\varpi} - \Xi(t) - [s_{n+1} - \tilde{s}_{n+1}]$$
$$= \varpi - s_{n+1} - [\tilde{\varpi} - \tilde{s}_{n+1}] - \Xi(t)$$
$$\leq 2\epsilon\Theta_1 + \Xi_m, t \geq t_{1\epsilon}. \tag{2.79}$$

The proof is completed.

Remark 2.6 The fault detection strategy (2.73) is an approach that uses the total uncertainty to realize fault diagnosis through the cancellation principle, which has significant advantages in incipient fault detection and diagnosis. If the fault amplitude is less than the disturbance remaining term, the fault detection strategy will fail. The overall structure will still maintain the PFTC mechanism, which is not available in most AFTC approaches. The traditional AFTC approaches are limited by the fault diagnosis accuracy. The proposed AFTC framework integrates the characteristics of PFTC and traditional AFTC, and fully guarantees the control performance.

2.4.2 Controller Design Techniques

The controller reconstruction rules are

$$u' = \begin{cases} -\bar{G}^{-1}(s_{n+1} + A_{0 \sim n-1} S), t < t_{th}, \\ -\bar{G}^{-1}(s_{n+1} + A_{0 \sim n-1} S + \hat{\delta}), t \geq t_{th}, \end{cases} \tag{2.80}$$

where t_{th} is the fault detection time and controller switch time.

Theorem 2.4 *For the real-time FAS* (2.6) *under the active fault-tolerant controller* (2.80), *if Assumptions 2.1 and 2.2 hold, there exist positive scalars such that*

1. The observation error and tracking error are ultimately uniformly bounded.
2. For any $\epsilon \in (0, \epsilon_0)$, it holds that

$$\begin{cases} \|x_i - s_i - x_{Ri}\| \leq \epsilon^{n+2-i} \Theta_{a1}, t \geq t_{1a\epsilon}, i \in [n], \\ \|\varpi - \hat{\varpi}\| \leq \epsilon \Theta_{a1}, t \geq t_{1a\epsilon}, \\ \|x_i - x_{Ri}\| \leq \epsilon \Theta_{a2}, t \geq t_{2a\epsilon}, i \in [n], \\ \|y - y_d\| \leq \epsilon \Theta_{a3}, t \geq t_{Ra\epsilon}. \end{cases} \tag{2.81}$$

Proof The proof is similar to Theorem 2.1. The main difference is the partial dynamics after controller switching

$$\begin{cases} \dot{s}_n = s_{n+1} + h_n(\vartheta) + \bar{G}u', \\ u' = -\bar{G}^{-1}(s_{n+1} + A_{0 \sim n-1} S + \hat{\delta}). \end{cases} \tag{2.82}$$

The remaining proof is omitted.

Remark 2.7 Many classical robust PFTC approaches [124, 142] will lead to the loss of control performance. In the proposed AFTC framework, disturbances rejection depends on the adaptive estimation, and fault tolerance depends on the switching control law. Therefore, it has higher tracking control precision and fault tolerance performance.

2.5   Simulation Studies and Results

To illustrate the effectiveness of the proposed AFTC framework, the following numerical FAS model is considered as

$$
\begin{cases}
x^{(3)} = F(x^{(0\sim2)}, \xi) + G(x^{(0\sim2)}, \xi)u + K(x^{(0\sim2)}, \xi)d, \\
y = x,
\end{cases}
\tag{2.83}
$$

where

$$
F(x^{(0\sim2)}, \xi) = 2\sin x^2 + \frac{\dot{x}}{1 + \dot{x}^2} + \tanh \ddot{x} + x\xi,
\tag{2.84}
$$

$$
G(x^{(0\sim2)}, \xi) = 5 + 0.1\sin t, \ K(x^{(0\sim2)}, \xi) = \sin x + \cos \dot{x} + 2 + \xi.
\tag{2.85}
$$

The unmodeled dynamics is set as $\xi = 0.2\sin t$, the process disturbance is set as $d = 5\sin t$, and the reference signal is designed as $y_d = 5\sin t + \cos t + 2$. The control task is to design the controller u so that y tracks y_d when the nonlinear dynamics is unknown.

2.5.1   Data Modeling Validation

Let $A_{0\sim2} = [2, 4, 3], \bar{G} = 5, \epsilon = 0.002, \mathfrak{a} = 5$, the observer and controller integration is designed as

$$
\begin{cases}
\dot{\hat{x}}_1 = \hat{x}_2 + 4\Gamma^{-1}(t)\left(y - \hat{x}_1\right), \\
\dot{\hat{x}}_2 = \hat{x}_3 + 7\Gamma^{-2}(t)\left(y - \hat{x}_1\right), \\
\dot{\hat{x}}_3 = \hat{\omega} + 6\Gamma^{-3}(t)\left(y - \hat{x}_1\right) + \bar{G}u, \\
\dot{\hat{\omega}} = 2\Gamma^{-4}(t)\left(y - \hat{x}_1\right), \\
u = -\bar{G}^{-1}\left[\hat{\omega} - x_{R4} + A^{0\sim2}(\hat{X} - X_R)\right].
\end{cases}
\tag{2.86}
$$

All initial system values are set as zero, and the above observation and tracking control results are shown in Fig. 2.1. The results display that the state and total uncertainty of (2.83) are well estimated, and the output tracking control task is also completed, which illustrating Theorem 2.1.

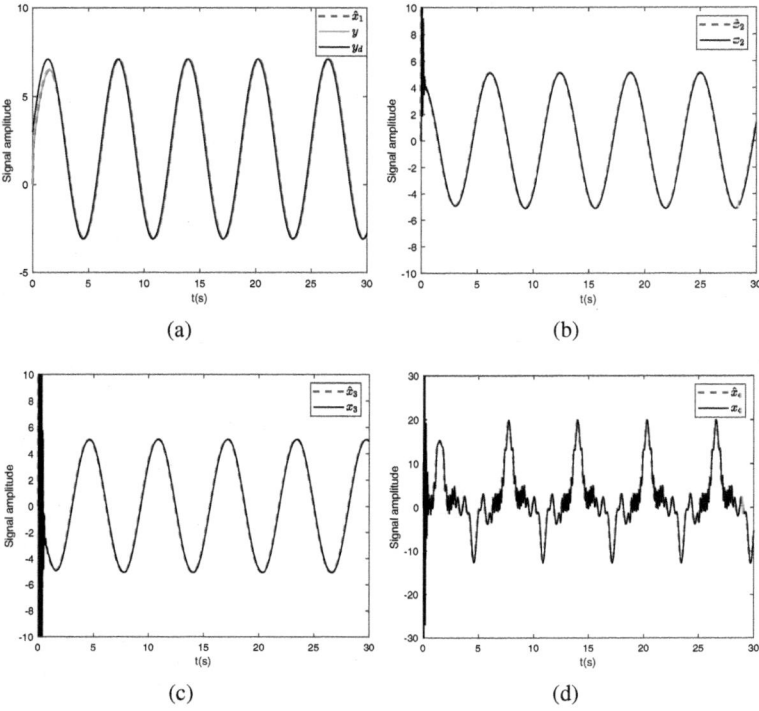

Fig. 2.1 The signal trajectories in the dynamic data modeling experiment

The corresponding dynamic data model is

$$
\begin{cases}
\dot{S} = \Phi_A S + \mathfrak{D}_1, \mathfrak{D}_1 = \begin{bmatrix} 4\Gamma^{-1}(t)(y - s_1 - x_{R1}) \\ 7\Gamma^{-2}(t)(y - s_1 - x_{R1}) \\ 6\Gamma^{-3}(t)(y - s_1 - x_{R1}) \end{bmatrix}, \\
\dot{s}_4 = 2\Gamma^{-4}(t)(y - s_1 - x_{R1}) - \dot{x}_{R4}.
\end{cases}
\tag{2.87}
$$

As can be seen from Fig. 2.2, the control performance based on the data model is consistent with the control performance based on the model-driven controller. Meanwhile, the response speed of (2.87) is faster than that of the model-driven system. The reason is that the model-driven controller is robust, while the proposed dynamic data modeling method is adaptive. The response speed of robust control is slower than that of adaptive control.

Fig. 2.2 The dynamic data
modeling experiment

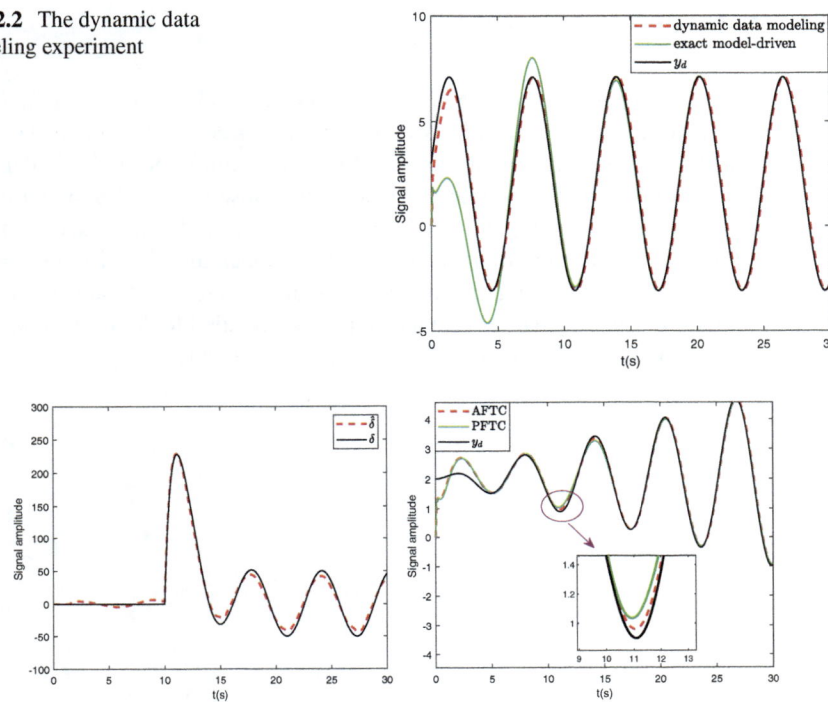

Fig. 2.3 The signal trajectories in the FTC experiment

2.5.2   Control System Validation

The controller in Sect. 2.3 is a PFTC strategy because it adaptively compensates
the total uncertainty including faults. PFTC plays an important role in early fault
compensation, but when the system encounters a more serious fault, PFTC often can
not complete the fault compensation in time. In order to illustrate the performance
of AFTC in Sect. 2.4, a component fault model is considered as

$$\delta = \begin{cases} 0, t < 10\text{s}, \\ 500(t-10)\exp(10-t) + 50\sin(t-10), t \geq 10\text{s}, \end{cases} \tag{2.88}$$

The component fault has a certain degree of severity due to the high amplitude. To
increase the difficulty of tracking control, the reference signal is designed as $y_d = 0.1t\sin t + 2$. As shown in Fig. 2.3, the AFTC compensates faults faster than the
PFTC, and the AFTC has better control performance when faults occur. Especially
for novel faults, the proposed AFTC has greater application value.

2.6 Notes and References

This chapter proposes a novel AFTC framework for a class of uncertain FASs with component faults. The main highlights are: (1) Based on the output data, the observer and controller integration design of the uncertain FAS model is derived, and the advantages of FAS theory are revealed—the closed-loop FAS model meets the generalized duality principle. (2) The dynamic data modeling structure of the FAS model restores the important features hidden under the closed-loop control, which is conducive to analyzing the uncertainties, disturbances and faults of closed-loop systems. (3) The original PFTC mechanism is extended to AFTC mechanism, which ensures the applicability of FAS theory in fault tolerance.

Chapter 3
Robust Adaptive FTC for FASs Against Actuator Faults

This chapter develops a robust adaptive fault-tolerant control (RAFTC) methodology for nonlinear high-order fully actuated systems (FASs) subject to concurrent multiplicative and additive actuator faults. Unlike conventional state-space approaches, the proposed FAS approaches enables global stabilization through tailored nonlinear control laws while preserving the inherent full-actuation and decoupled characteristics of physical systems. Initiating from the tracking error dynamics in FAS framework, a tracking differentiator (TD) is engineered to computationally resolve reference signals and their temporal derivatives. The study pioneers the integration of an extended state observer within the FAS structure, significantly broadening the theoretical applicability boundaries. A Lyapunov-based fault-compensation control scheme is systematically constructed, guaranteeing uniformly ultimately bounded tracking error stability. Comprehensive benchmark simulation studies demonstrate the superior performance against existing approaches, particularly in fault accommodation scenarios.

3.1 Overview and Challenges

The component fault in the previous chapter is an additive fault type, while the common additive deviation and multiplicative gain-attenuated actuator faults are more demanding fault types to handle. The fault-tolerant control (FTC) technology for actuator faults based on first-order state-space models has achieved a large number of results [6]. Under unreliable network links, an adaptive FTC technology based on high-gain observers can compensate actuator faults in multi-agent systems [73]. For a class of Markov stochastic systems, the backstepping technique can effectively handle additive and multiplicative actuator faults [111]. The existing first-order state space methods of FTC still have the characteristic of complex structure.

© The Author(s) 2026
D. Zhou, M. Cai, *Fault-Tolerant Control for Fully Actuated Systems*,
https://doi.org/10.1007/978-981-95-0691-0_3

Facing more general actuator faults, Chap. 3 combines the extended state observer technology and the FAS structure to design a standard RAFTC framework. For multiplicative actuator faults, a robust FTC strategy is adopted, and for additive actuator faults, an adaptive FTC strategy is utilized. This framework is easy to tune parameters and has good fault tolerance performance.

3.2 Problem Definition and Preliminaries

Consider a class of FAS models with actuator faults and process disturbances

$$
\begin{cases}
x^{(n)} = F(x^{(0 \sim n-1)}) + G(x^{(0 \sim n-1)})\tilde{u} + K(x^{(0 \sim n-1)})d, \\
y = x + v,
\end{cases}
\tag{3.1}
$$

where $x \in \mathbb{R}^m, \tilde{u} \in \mathbb{R}^m, d \in \mathbb{R}^r, y \in \mathbb{R}^m, v \in \mathbb{R}^m$ are the basic state, the control input with possible actuator faults, the process disturbances, the measurement output and the measurement noise, respectively. Functions $F(\cdot) \in \mathbb{R}^m, G(\cdot) \in \mathbb{R}^{m \times m}, K(\cdot) \in \mathbb{R}^{m \times r}$ are known, and the noise is tiny. The actuator fault is modeled as

$$
\tilde{u} = (1 - \rho)u + u_f,
\tag{3.2}
$$

where $\rho \in (0, 1)$ is a bounded and differentiable gain decay fault, u_f is a bounded and differentiable deviation fault, and u is the actual control signal.

Assumption 3.1 ([20]) The system (3.1) is fully actuated, i.e.,

$$
\|G(x^{(0 \sim n-1)})\| \neq 0, \forall x^{(i-1)} \in \mathbb{R}^m, i \in [n].
\tag{3.3}
$$

Assumption 3.2 ([44, 133]) From a scalar perspective, the system state $x_j^{(i-1)}, i \in [n]$, the controller $u_j, j \in [m]$, and the disturbance $d_k, \dot{d}_k, k \in [r]$ are bounded. Meanwhile, the partial derivatives of $F(\cdot), G(\cdot), K(\cdot)$ are all bounded. In addition, the disturbance function $K(\cdot)d$ has a known bound D.

Remark 3.1 Assumption 3.1 postulates bounded system signals with associated nonlinearities satisfying Lipschitz continuity conditions, which aligns with conventional frameworks in the FTC research [82]. This assumption holds practical validity across prevalent engineering applications including multi-joint robotic manipulators and permanent magnet synchronous motors where inherent physical constraints naturally enforce such bounded-input bounded-state characteristics.

Assumption 3.3 ([26]) A pre-estimator $u_{f0} = u_{f0}(t) \in \mathbb{R}^m$ is designed for the time-varying fault u_f, which meets $\|u_f(t) - u_{f0}(t)\| \leq \delta_1, \|\dot{u}_f(t) - \dot{u}_{f0}(t)\| \leq \delta_2, \forall t \geq 0$ with $\delta_1, \delta_2 > 0$.

Remark 3.2 The fault pre-estimator is effective, because constant and slowly time-varying faults can be pre-estimated. Especially under engineering experience, some system operation data will produce some information. The chapter do not demand the accuracy of the pre-estimator, but only need some useful information from the pre-estimator.

The control task is to let x track the smooth reference signal x_d. Define the tracking error

$$e = x - x_d. \tag{3.4}$$

The high-order derivatives of (3.4) are

$$e^{(i)} = x^{(i)} - x_d^{(i)}, i \in [n]. \tag{3.5}$$

It is reasonable that

$$F \triangleq F(x^{(0 \sim n-1)}) = F(e^{(0 \sim n-1)} + x_d^{(0 \sim n-1)}), \tag{3.6}$$

$$G \triangleq G(x^{(0 \sim n-1)}) = G(e^{(0 \sim n-1)} + x_d^{(0 \sim n-1)}), \tag{3.7}$$

$$K \triangleq K(e^{(0 \sim n-1)} + x_d^{(0 \sim n-1)}). \tag{3.8}$$

Then the tracking error dynamics is

$$e^{(n)} = F + G\left((1 - \rho)u + u_f\right) + Kd - x_d^{(n)}. \tag{3.9}$$

It follows that (3.4)–(3.9) require the derivatives of x_d. This chapter chooses a TD to numerically solve the derivatives of x_d. The TD can reduce the error between the initial system state and the initial reference signal. It is supposed that the entries of $f : \mathbb{R}^{(n+1)m} \to \mathbb{R}^m$ is locally Lipschitz, and the asymptotic equilibrium point of the following system (3.10) is $[0^\mathsf{T}, 0^\mathsf{T}, \cdots, 0^\mathsf{T}]^\mathsf{T}$.

$$\begin{cases} \dot{x}_1 = x_2, \\ \dot{x}_2 = x_3 \\ \quad \vdots \\ \dot{x}_n = x_{n+1}, \\ \dot{x}_{n+1} = f(x_1, x_2, \cdots, x_{n+1}), \end{cases} \tag{3.10}$$

When the reference signal x_d is smooth and its first-order derivative are bounded, the TD is designed as

$$
\begin{cases}
\dot{x}_{R1} = x_{R2}, \\
\dot{x}_{R2} = x_{R3}, \\
\quad \vdots \\
\dot{x}_{Rn} = x_{R(n+1)}, \\
\dot{x}_{R(n+1)} = R^n f\left(x_{R1} - x_d, \frac{x_{R2}}{R}, \cdots, \frac{x_{R(n+1)}}{R^n}\right),
\end{cases}
\tag{3.11}
$$

where $X_R = [x_{R1}^{\mathsf{T}}, x_{R2}^{\mathsf{T}}, \cdots, x_{Rn}^{\mathsf{T}}]^{\mathsf{T}}$ and R represents the convergence parameter.

Lemma 3.1 ([43]) *Given any initial value and any positive constant $\mathfrak{a}$, x_{R1} uniformly converges to x_d on the interval $[\mathfrak{a}, \infty)$ when $R \to \infty$. Furthermore, $x_{Ri}, i \in [n]$ approximates $x_d^{(i)}, i \in [n]$ in the generalized weak convergence sense.*

In fact, the TD is an auxiliary signal generator to track the reference signal slowly. Based on Lemma 3.1, the derivatives $x_d^{(i-1)}, i \in [n+1]$ in (3.4)–(3.9) can be reasonably replaced by x_{Ri}. To facilitate the FAS controller design, there are two lemmas.

Lemma 3.2 *There are some variable definitions $\tilde{u}_f = u_f - \hat{u}_f, \tilde{u}_{e0} = \hat{u}_f - u_{f0}, \tilde{u}_{r0} = u_f - u_{f0}$ where $\hat{u}_f$ is the estimation of u_f. The following conclusions hold:*

$$
\tilde{u}_{r0}^{\mathsf{T}} \tilde{u}_f \le \frac{1}{2}(\delta_1^2 + \|\tilde{u}_f\|^2), \dot{\tilde{u}}_{r0}^{\mathsf{T}} \tilde{u}_f \le \frac{1}{2}(\delta_2^2 + \|\tilde{u}_f\|^2).
\tag{3.12}
$$

Proof Based on Assumption 3.3 and Young's inequation, it turns out that

$$
\tilde{u}_{r0}^{\mathsf{T}} \tilde{u}_f \le \frac{1}{2}(\|u_f - u_{f0}\|^2 + \|\tilde{u}_f\|^2) \le \frac{1}{2}(\delta_1^2 + \|\tilde{u}_f\|^2),
\tag{3.13}
$$

$$
\dot{\tilde{u}}_{r0}^{\mathsf{T}} \tilde{u}_f \le \frac{1}{2}(\|\dot{u}_f - \dot{u}_{f0}\|^2 + \|\tilde{u}_f\|^2) \le \frac{1}{2}(\delta_2^2 + \|\tilde{u}_f\|^2).
\tag{3.14}
$$

Lemma 3.3 ([26]) *For any positive constant μ, there exists a set of matrices $A_i \in R^{m \times m}, i = 0, 1, \cdots, n-1$ satisfying*

$$
\Re \lambda_j (\Phi_A) < -\frac{\mu}{2}, j \in [nm].
\tag{3.15}
$$

As a result, there exists a positive definite matrix $P = [P_1, P_2, \cdots, P_n], P_i \in \mathbb{R}^{nm \times m}$ satisfying

$$
\Phi_A^{\mathsf{T}} P + P \Phi_A \le -\mu P.
\tag{3.16}
$$

3.3 Design of Fault-Tolerant Control Systems

There are two tasks in the FTC design. One is disturbance rejection and the other is fault tolerance. At first, the corresponding observer needs to be designed in that the full-order states are not directly measured.

3.3.1 Extended State Observer Design

In most conditions, the exact system state cannot be always available. However, the FAS controller structure relies on the exact system state. Thus, an extended state observer (ESO) for the tracking error system (3.9) is designed as below.

Let $[z_1^\mathsf{T}, z_2^\mathsf{T}, \cdots, z_n^\mathsf{T}]^\mathsf{T} = \hat{e}^{(0 \sim n-1)}$ where $\hat{e}^{(i)}$ are the estimations of $e^{(i)}$ and z_{n+1} is the estimation of $F + Kd - x_d^{(n)}$, and then a nonlinear ESO is designed as

$$
\begin{cases}
\dot{z}_1 = z_2 - h_1(z_1 - y + x_d), \\
\quad\vdots \\
\dot{z}_n = z_{n+1} - h_n(z_1 - y + x_d) + Gu, \\
\dot{z}_{n+1} = -h_{n+1}(z_1(t) - y + x_d),
\end{cases}
\tag{3.17}
$$

where $h_i(\cdot), i \in [n+1]$ are the nonlinear functions to be designed. A nonlinear function is considered as

$$
\mathrm{fal}(\alpha, \varepsilon) =
\begin{cases}
\dfrac{z_1 - y + x_d}{\varepsilon^{1-\alpha}}, & \|z_1 - y + x_d\| \le \varepsilon, \\
\|z_1 - y + x_d\|^\alpha \mathrm{sign}(z_1 - y + x_d), & \|z_1 - y + x_d\| > \varepsilon,
\end{cases}
\tag{3.18}
$$

where α, ε are positive constants to be designed. Then, the nonlinear function $h_i(\cdot), \in [n+1]$ are designed as

$$
\begin{cases}
h_1(z_1 - y + x_d) = \mathscr{H}_1 \mathrm{fal}(\alpha, \varepsilon), \\
h_2(z_1 - y + x_d) = \mathscr{H}_2 \mathrm{fal}(\alpha, \varepsilon), \\
\quad\vdots \\
h_{n+1}(z_1 - y + x_d) = \mathscr{H}_{n+1} \mathrm{fal}(\alpha, \varepsilon),
\end{cases}
\tag{3.19}
$$

where $\mathscr{H}_i, i \in [n+1]$ are the designed observer gain. There is sufficient evidence on the convergence of ESO in [82].

Remark 3.3 Compared with the traditional ESO without the sign function, the proposed ESO can show good noise suppression performance because the parameter $\alpha \in (0, 1)$ can be adjusted to reduce noise influence. However, the chattering phenomenon becomes an inevitable problem due to the sign function. Fortunately, the chattering degree is positively correlated with α and negatively correlated with ε. Besides, the traditional ESO and the proposed ESO can be jointly utilized, which can not only suppress noise influence, but also weaken the chattering phenomenon.

Lemma 3.4 ([44]) *If there exist positive constants $C_i, i \in [5]$ and continuous positive definite functions $\mathscr{V}, \mathscr{W} : \mathbb{R}^{(n+1)m} \to \mathbb{R}$ satisfying*

1. $C_1\|z\|^2 \le \mathscr{V}(z) \le C_2\|z\|^2, C_3\|z\|^2 \le \mathscr{W}(z) \le C_4\|z\|^2,$
2. $\sum\limits_{i=1}^{n}[\dfrac{\partial \mathscr{V}}{\partial z_i}]^{\mathsf{T}}(z_{i+1} - h_i(z_1)) - [\dfrac{\partial \mathscr{V}}{\partial z_{n+1}}]^{\mathsf{T}}h_{n+1}(z_1) \le -\mathscr{W}(z),$
3. $\left\|\dfrac{\partial \mathscr{V}}{\partial z_{n+1}}\right\| \le C_5\|z\|,$

where $z = [z_1^{\mathsf{T}}, z_2^{\mathsf{T}}, \cdots, z_{n+1}^{\mathsf{T}}]^{\mathsf{T}}$. It follows that

$$\lim_{\mathscr{H}_i \to \infty, t \to \infty} \|z_i(t) - e^{(i-1)}(t)\| \le \sigma, i \in [n+1], \tag{3.20}$$

where σ is related to noise amplitude.

In fact, the bound σ is a small constant because the noise in the FAS model is tiny. Assuming that the original FAS model (3.1) meets the separability principle, an observer-based FAS controller can be designed.

3.3.2 Robust Controller Design

In order to realize FTC, the RAFTC framework is divided into three items:

$$u = -G^{-1}(\hat{e}^{(0\sim n-1)} + X_R)(u_0 + u_1 + u_2), \tag{3.21}$$

where the control law u_0 can transform the high-order FAS into a linear time-invariant system, and u_1, u_2 can cope with faults and disturbances. The three items are reasonably assumed to be bounded. The specific forms of the three items are as below

$$P_L \triangleq P\begin{bmatrix} \mathbf{0}_{(n-1)m} \\ I_m \end{bmatrix} = P_n, \tag{3.22}$$

$$u_0 = A_{0\sim n-1}\hat{e}^{(0\sim n-1)} + F(\hat{e}^{(0\sim n-1)} + X_R) - x_{R(n+1)}, \tag{3.23}$$

$$u_1(t) = \frac{D^2}{4\epsilon}P_L^{\mathsf{T}}\hat{e}^{(0\sim n-1)}, \tag{3.24}$$

$$\begin{cases} \dot{\hat{u}}_f = G^{\mathsf{T}}(\hat{e}^{(0 \sim n-1)} + X_R)P_L^{\mathsf{T}}\hat{e}^{(0 \sim n-1)} - (\mu + 1)(\hat{u}_f - u_{f0}) + \dot{u}_{f0}, \\ u_2 = G(\hat{e}^{(0 \sim n-1)} + X_R)\hat{u}_f, \end{cases}$$

$$(3.25)$$

where $\epsilon > 0$ is a pre-designed parameter.

Theorem 3.1 *For the tracking error system (3.9) with a pre-estimator u_{f0}, let the selected matrices $A_{i-1} \in \mathbb{R}^{m \times m}, \in [n]$ satisfy (3.15), and then the tracking error and the bias fault estimation error will be uniformly ultimately bounded under the RAFTC framework (3.21)-(3.25). The error bound can be described as follows*

$$\Xi_{\mu,\epsilon,\theta}(0) = \left\{ \begin{bmatrix} e^{(0 \sim n-1)} \\ \tilde{u}_f \end{bmatrix} \middle| \left\| e^{(0 \sim n-1)} \right\|_P^2 + \|\tilde{u}_f\|^2 \le 2\frac{\epsilon_m + \theta}{\mu} \right\},$$

$$(3.26)$$

where

$$\left\| e^{(0 \sim n-1)} \right\|_P^2 = (e^{(0 \sim n-1)})^{\mathsf{T}} P e^{(0 \sim n-1)},$$

$$(3.27)$$

$$\epsilon_m = \epsilon \frac{(D + u_m)^2}{D^2},$$

$$(3.28)$$

$$\theta = \frac{1}{2}\left[\delta_2^2 + (\mu + 1)\delta_1^2\right],$$

$$(3.29)$$

u_m *represents a positive constant from (3.40).*

Proof The tracking error system under the RAFTC framework (3.21)–(3.25) is

$$e^{(n)} + A_{0 \sim n-1}\hat{e}^{(0 \sim n-1)} = \psi_1(\hat{e}^{(0 \sim n-1)}) + \psi_2(\hat{e}^{(0 \sim n-1)}).$$

$$(3.30)$$

Assuming that the observation error of ESO is very tiny and the above nonlinear functions are Lipschitz, the overall observation error can be regarded as a small disturbance ϖ. The closed-loop error system (3.30) can be written as

$$e^{(n)} + A_{0 \sim n-1}e^{(0 \sim n-1)} = \psi_1(e^{(0 \sim n-1)}) + \psi_2(e^{(0 \sim n-1)}),$$

$$(3.31)$$

where

$$\psi_1 \triangleq \psi_1(e^{(0 \sim n-1)}) = \rho(u_0 + u_1 + u_2) + Kd + \varpi - \frac{D^2}{4\epsilon}P_L^{\mathsf{T}}e^{(0 \sim n-1)},$$

$$(3.32)$$

$$\psi_2 \triangleq \psi_2(e^{(0 \sim n-1)}) = G\tilde{u}_f.$$

$$(3.33)$$

Moreover, (3.31) can be expressed as

$$\dot{e}^{(0\sim n-1)} = \Phi_A e^{(0\sim n-1)} + \begin{bmatrix} \mathbf{0}_{(n-1)m} \\ \psi_1 + \psi_2 \end{bmatrix}, \tag{3.34}$$

Based on Lemma 3.3, there exists a positive definite matrix P satisfying (3.16). Consider the Lyapunov function

$$\Theta = \frac{1}{2}(e^{(0\sim n-1)})^\mathsf{T} P e^{(0\sim n-1)} + \frac{1}{2}\tilde{u}_f^\mathsf{T}\tilde{u}_f. \tag{3.35}$$

The derivative of (3.35) is

$$\begin{aligned}
\dot{\Theta} &= \frac{1}{2}\left(\dot{e}^{(0\sim n-1)}\right)^\mathsf{T} P e^{(0\sim n-1)} + \frac{1}{2}\left(e^{(0\sim n-1)}\right)^\mathsf{T} P\dot{e}^{(0\sim n-1)} + \dot{\tilde{u}}_f^\mathsf{T}\tilde{u}_f \\
&= \frac{1}{2}\left(\Phi e^{(0\sim n-1)} + \begin{bmatrix} \mathbf{0}_{(n-1)m} \\ \psi_1 + \psi_2 \end{bmatrix}\right)^\mathsf{T} P e^{(0\sim n-1)} + \dot{\tilde{u}}_f^\mathsf{T}\tilde{u}_f \\
&\quad + \frac{1}{2}\left(e^{(0\sim n-1)}\right)^\mathsf{T} P\left(\Phi_A e^{(0\sim n-1)} + \begin{bmatrix} \mathbf{0}_{(n-1)m} \\ \psi_1 + \psi_2 \end{bmatrix}\right) \\
&= \frac{1}{2}\left(e^{(0\sim n-1)}\right)^\mathsf{T}\left(\Phi_A^\mathsf{T} P + P\Phi_A\right)e^{(0\sim n-1)} \\
&\quad + \left(e^{(0\sim n-1)}\right)^\mathsf{T} P\begin{bmatrix} \mathbf{0}_{(n-1)m} \\ \psi_1 + \psi_2 \end{bmatrix} + \dot{\tilde{u}}_f^\mathsf{T}\tilde{u}_f \\
&\leq -\frac{\mu}{2}\left(e^{(0\sim n-1)}\right)^\mathsf{T} P e^{(0\sim n-1)} + \left(e^{(0\sim n-1)}\right)^\mathsf{T} P_L(\psi_1 + \psi_2) + \dot{\tilde{u}}_f^\mathsf{T}\tilde{u}_f \\
&= -\mu\Theta + \Theta_1 + \Theta_2, \tag{3.36}
\end{aligned}$$

where

$$\Theta_1 = \left(e^{(0\sim n-1)}\right)^\mathsf{T} P_L\psi_1, \tag{3.37}$$

$$\Theta_2 = \left(e^{(0\sim n-1)}\right)^\mathsf{T} P_L\psi_2 + \dot{\tilde{u}}_f^\mathsf{T}\tilde{u}_f + \frac{\mu}{2}\tilde{u}_f^\mathsf{T}\tilde{u}_f. \tag{3.38}$$

Based on (3.24) and (3.32), it turns out that

$$\begin{aligned}
\Theta_1 &= \left(e^{(0\sim n-1)}\right)^\mathsf{T} P_L\psi_1 \\
&= -\frac{D^2}{4\epsilon}\left[\left(e^{(0\sim n-1)}\right)^\mathsf{T} P_L P_L^\mathsf{T} e^{(0\sim n-1)}\right] \\
&\quad + \left(e^{(0\sim n-1)}\right)^\mathsf{T} P_L\left[\rho(u_0 + u_1 + u_2) + Kd + \varpi\right]
\end{aligned}$$

$$\leq -\frac{D^2}{4\epsilon} \left\| P_L^{\mathsf{T}} e^{(0\sim n-1)} \right\|^2 + \left\| \rho(u_0 + u_1 + u_2) + Kd + \varpi \right\| \left\| P_L^{\mathsf{T}} e^{(0\sim n-1)} \right\|.$$

$$(3.39)$$

Given the FAS controller u and the disturbance item ϖ are bounded, there exists a positive constant u_m satisfying

$$\| \rho(u_0 + u_1 + u_2) + \varpi \| \leq_m .$$

$$(3.40)$$

Then, (3.39) satisfies

$$\Theta_1 \leq -\frac{D^2}{4\epsilon} \left\| P_L^{\mathsf{T}} e^{(0\sim n-1)} \right\|^2 + (D + u_m) \left\| P_L^{\mathsf{T}} e^{(0\sim n-1)} \right\|$$

$$\leq \epsilon \frac{(D + u_m)^2}{D^2} = \epsilon_m.$$

$$(3.41)$$

Based on Lemma 3.2 and (3.33), it follows that

$$\Theta_2 = \left(e^{(0\sim n-1)} \right)^{\mathsf{T}} P_L \psi_2 + \dot{\tilde{u}}_f^{\mathsf{T}} \tilde{u}_f + \frac{\mu}{2} \tilde{u}_f^{\mathsf{T}} \tilde{u}_f$$

$$= \left[\left(e^{(0\sim n-1)} \right)^{\mathsf{T}} P_L G + \dot{\tilde{u}}_f^{\mathsf{T}} \right] \tilde{u}_f + \frac{\mu}{2} \tilde{u}_f^{\mathsf{T}} \tilde{u}_f.$$

$$(3.42)$$

Lemma 3.2 also yields

$$\tilde{u}_{r0} - \tilde{u}_{e0} - \tilde{u}_f = 0.$$

$$(3.43)$$

Substituting (3.43) into (3.42) yields

$$\Theta_2 = \left[\left(e^{(0\sim n-1)} \right)^{\mathsf{T}} P_L G + \dot{\tilde{u}}_f^{\mathsf{T}} \right] \tilde{u}_f + \frac{\mu}{2} \tilde{u}_f^{\mathsf{T}} \tilde{u}_f$$

$$+ (\mu + 1) \left(\tilde{u}_{r0} - \tilde{u}_{e0} - \tilde{u}_f \right)^{\mathsf{T}} \tilde{u}_f$$

$$= \left[\left(e^{(0\sim n-1)} \right)^{\mathsf{T}} P_L G - \dot{\tilde{u}}_{e0}^{\mathsf{T}} + \dot{\tilde{u}}_{r0}^{\mathsf{T}} \right] \tilde{u}_f$$

$$+ (\mu + 1) \left(-\tilde{u}_{e0} + \tilde{u}_{r0} - \tilde{u}_f \right)^{\mathsf{T}} \tilde{u}_f + \frac{\mu}{2} \tilde{u}_f^{\mathsf{T}} \tilde{u}_f$$

$$= \left[\left(e^{(0\sim n-1)} \right)^{\mathsf{T}} P_L G - \dot{\tilde{u}}_{e0}^{\mathsf{T}} \right] \tilde{u}_f + \dot{\tilde{u}}_{r0}^{\mathsf{T}} \tilde{u}_f + \frac{\mu}{2} \tilde{u}_f^{\mathsf{T}} \tilde{u}_f$$

$$- (\mu + 1)\tilde{u}_{e0}^{\mathsf{T}} \tilde{u}_f + (\mu + 1) \left(\tilde{u}_{r0}^{\mathsf{T}} \tilde{u}_f - \tilde{u}_f^{\mathsf{T}} \tilde{u}_f \right)$$

$$= \left[\left(e^{(0\sim n-1)} \right)^{\mathsf{T}} P_L G - \dot{\tilde{u}}_{e0}^{\mathsf{T}} - (\mu + 1)\tilde{u}_{e0}^{\mathsf{T}} \right] \tilde{u}_f + \dot{\tilde{u}}_{r0}^{\mathsf{T}} \tilde{u}_f$$

$$+ (\mu + 1)\tilde{u}_{r0}^{\mathsf{T}} \tilde{u}_f - \frac{\mu + 2}{2} \tilde{u}_f^{\mathsf{T}} \tilde{u}_f.$$

$$(3.44)$$

It can be obtained that

$$\dot{\tilde{u}}_{e0}^{\mathsf{T}} + (\mu + 1)\tilde{u}_{e0}^{\mathsf{T}} = \dot{\hat{u}}_{f}^{\mathsf{T}} - \dot{u}_{f0}^{\mathsf{T}} + (\mu + 1)(\hat{u}_{f} - u_{f0}). \tag{3.45}$$

Based on (3.25) and (3.45), Θ_2 satisfies

$$
\begin{aligned}
\Theta_2 &= \dot{\tilde{u}}_{r0}^{\mathsf{T}}\tilde{u}_f + (\mu + 1)\tilde{u}_{r0}^{\mathsf{T}}\tilde{u}_f - \frac{\mu + 2}{2}\tilde{u}_f^{\mathsf{T}}\tilde{u}_f \\
&\leq \frac{1}{2}(\delta_2^2 + \|\tilde{u}_f\|^2) + \frac{\mu + 1}{2}(\delta_1^2 + \|\tilde{u}_f\|^2) - \frac{\mu + 2}{2}\|\tilde{u}_f\|^2 \\
&= \frac{1}{2}\left[\delta_2^2 + (\mu + 1)\delta_1^2\right] = \theta.
\end{aligned}
\tag{3.46}
$$

Combining (3.36), (3.39) and (3.46) yields

$$\dot{\Theta} \leq \mu\Theta + \epsilon_m + \theta. \tag{3.47}$$

It can be concluded that

$$
\begin{aligned}
\Theta &\leq \Theta(0)e^{-\mu t} + \frac{\epsilon_m + \theta}{\mu}(1 - e^{-\mu t}) \\
&= \left[\Theta(0) - \frac{\epsilon_m + \theta}{\mu}\right]e^{-\mu t} + \frac{\epsilon_m + \theta}{\mu} \\
&\to \frac{\epsilon_m + \theta}{\mu} \quad (t \to \infty).
\end{aligned}
\tag{3.48}
$$

Obviously, the appropriate parameters ϵ and μ will make the overall error as small as possible. The proof is completed.

In the closed-loop linearization process, various parametric matrices will lead to various control performance. The parametric solutions of linear matrices $A_{i-1}, i \in [n]$ and Φ_A are given as follows.

Lemma 3.5 ([23]) *For an arbitrarily chosen matrix $E \in \mathbb{R}^{nm \times nm}$ and a nonsingular matrix $V \in \mathbb{R}^{nm \times nm}$, the designed matrix $A_{0 \sim n-1}$ satisfies*

$$\Phi_A = VEV^{-1},$$

$$A_{0 \sim n-1} = -ZE^n V^{-1}(Z, E),$$

$$
V(Z, E) =
\begin{bmatrix}
Z \\
ZE \\
\vdots \\
ZE^{n-1}
\end{bmatrix},
$$

where $Z \in \mathbb{R}^{m \times nm}$ is an arbitrarily chosen matrix satisfying $|V(Z, E)| \neq 0$.

In practical solutions, E is usually a diagonalized matrix with negative diagonal elements. If the FAS model demands that E has the complex eigenvalues, the diagonal block $[-a, -b; b, -a]$ can be applied into E where $a, b > 0$.

The featured FAS architecture demonstrates dual methodological merits: fundamentally, it enables systematic linearization of nonlinear control problems through state-space reconfiguration techniques, effectively reducing complex nonlinear stabilization challenges to tractable linear equivalent formulations. Simultaneously, its explicit parametric decomposition empowers computationally efficient controller synthesis with enhanced design flexibility—the modular parameterization permits task-specific adaptation of control laws through gain-scheduling techniques without architectural modifications.

3.4 Simulation and Case Studies

In this section, a numerical example is presented to show the effectiveness of the proposed RAFTC framework. A nonlinear HOFAS model is considered as

$$
\begin{aligned}
\dddot{s} &= F(s, \dot{s}, \ddot{s}) + \tilde{u} + d, \\
y &= s + v,
\end{aligned}
\tag{3.49}
$$

where

$$
F(s, \dot{s}, \ddot{s}) = -(1 + \frac{\cos t}{2})\dot{s} - (1 + \sin \frac{t}{3})\ddot{s} + \sin \ddot{s}^2.
\tag{3.50}
$$

The disturbance is $d = 1 + 0.2 \sin(10t)$ and the noise is $v(t) = 0.01 \sin(10t)$. At $t = 25$s, the actuator fails in the following form:

$$
\tilde{u} = (\rho_1 + \rho_2 e^{-t})u + u_f,
$$

$$
\rho_1 = \begin{cases} 1, t \le 25\text{s}, \\ 0.7, t > 25\text{s}, \end{cases} \quad
\rho_2 = \begin{cases} 0, t \le 25\text{s}, \\ 0.3, t > 25\text{s}, \end{cases}
\tag{3.51}
$$

$$
u_f = \begin{cases} 0, t \le 25\text{s}, \\ 25 + 25e^{-t} \sin(10t), t > 25\text{s}. \end{cases}
$$

IThe reference signal is $x_d = 5 \sin t$, and the TD is designed as

$$
\begin{cases}
\dot{x}_{R1} = x_{R2}, \\
\dot{x}_{R2} = x_{R3} \\
\dot{x}_{R3} = x_{R4}, \\
\dot{x}_{R4} = -R^3[x_{R1} - x_d] - 4R^3 x_{R2} - 4R^2 x_{R3} - 4R x_{R4}.
\end{cases}
\tag{3.52}
$$

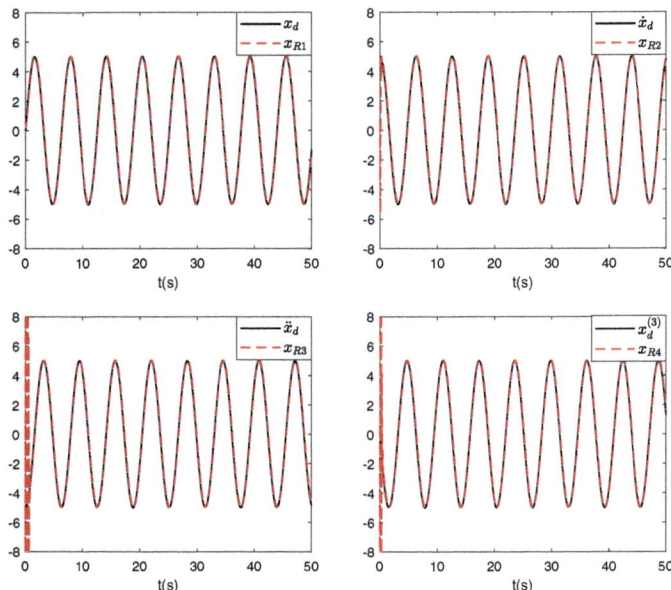

Fig. 3.1 The TD experiment

where R is set as 50 and the system sampling time is 0.001 s. From Fig. 3.1, the TD states can track the reference signal x_d and its derivatives quickly.

During the FAS parametric solution, the two matrices are selected as

$$E = \begin{bmatrix} -5 & -2 & 0 \\ 2 & -5 & 0 \\ 0 & 0 & -10 \end{bmatrix}, Z = \begin{bmatrix} 1 & 1 & 1 \end{bmatrix}. \tag{3.53}$$

Based on Lemma 3.5, it follows that $A_{0\sim2} = [290\ 129\ 20]$. A Lyapunov equation is

$$\left(\Phi_A + \frac{\mu}{2} I \right)^{\mathsf{T}} P + P \left(\Phi_A + \frac{\mu}{2} I \right) = -10^{-3} I, \tag{3.54}$$

where $\mu = 8$ satisfies Lemma 3.3. Solving (3.54) yields

$$P = \begin{bmatrix} 2200.5 & 532.23 & 30.354 \\ 532.23 & 149.58 & 8.7679 \\ 30.354 & 8.7679 & 0.5792 \end{bmatrix}. \tag{3.55}$$

In the observer design, the parameters are set as $\alpha = 0.5$, $\varepsilon = 1$ and the specific observer form is

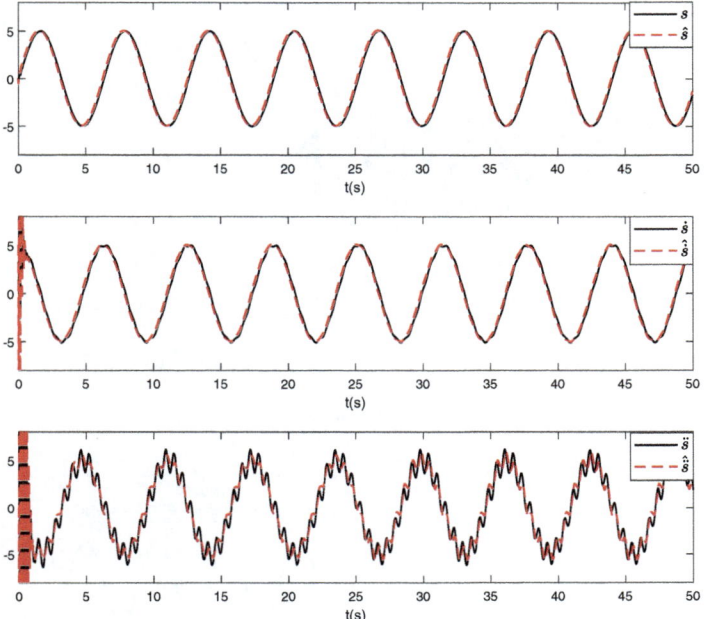

Fig. 3.2 The ESO experiment

$$\begin{cases} \dot{z}_1 = z_2 - 50 \times 4\mathrm{fal}(0.5, 1), \\ \dot{z}_2 = z_3 - 50^2 \times 7\mathrm{fal}(0.5, 1), \\ \dot{z}_3 = z_4 - 50^3 \times 6\mathrm{fal}(0.5, 1) + u, \\ \dot{z}_4 = -50^4 \times 2\mathrm{fal}(0.5, 1). \end{cases} \qquad (3.56)$$

The initial system state is $[0, 0, 0]^\mathsf{T}$ and the initial observation is $[0, 0, 0, 0]^\mathsf{T}$. The ESO experiment are as Fig. 3.2 shown. The observation error is bounded, indicating Lemma 3.4. In addition, it can be found that there is still weakened chattering phenomenon in the higher-order system state. And the reason for the weakened chattering phenomenon lies in the appropriate values of α, ε. The chattering degree is positively correlated with α and negatively correlated with ε.

Let $\epsilon = 0.1, D = 1.5, u_{f0} = 25$, the input signal and the fault estimation are shown in Fig. 3.3a, b. From Fig. 3.3a, b, the adaptive adjustment of the FAS controller occurrs at $t = 25$ s. The fault estimation error is ultimately bounded, indicating Theorem 3.1. The more accurate the pre-estimator, the better the fault estimation. In practical applications, a more appropriate pre-estimator can be chosen based on engineering experience, which can easily improve the control performance.

Finally, to further demonstrate the superiority of the proposed RAFTC frame-work, the classical active disturbance rejection control (ADRC) is selected for a

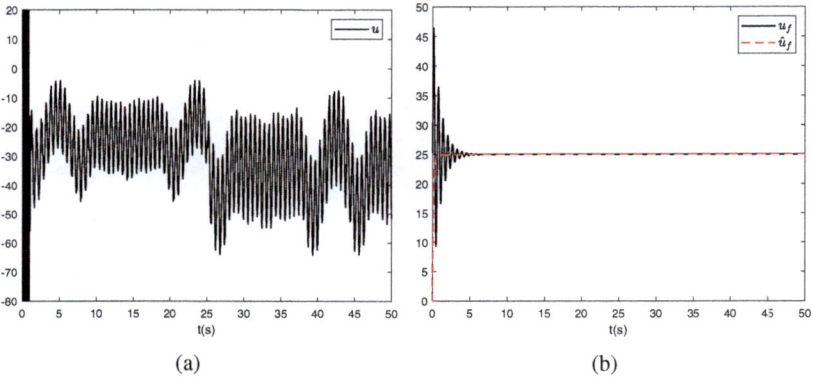

Fig. 3.3 The signal trajectories in RAFTC experiment

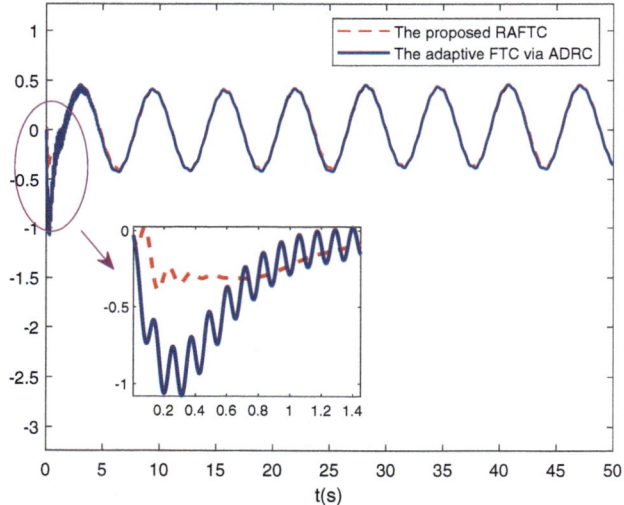

Fig. 3.4 The tracking error under the RAFTC and the adaptive FTC via ADRC

comparative experiment. Different from the proposed FAS structure, the ADRC structure is based on the state-space model. The ADRC method is an adaptive FTC technique, which can dynamically compensate actuator faults. In the comparative experiment, the reference signal is set as $x_d = 5 \sin t$. From Fig. 3.4, it is obvious that the tracking error under two approaches are bounded and the two error bounds depend on disturbances, faults and noise. Obviously, the proposed RAFTC approach has two advantages: (1) The proposed RAFTC framework can initially stabilize the FAS faster, and the system oscillation is much weaker. (2) The proposed RAFTC framework can yield smaller error, especially at the fault occurrence. As a result, the proposed RAFTC framework for the high-order FASs has a certain application prospect.

3.5 Notes and References

In this chapter, a RAFTC framework is proposed for FASs to cope with process disturbances and actuator faults. The main highlights of the proposed approach contain: (1) The FAS model is a control-oriented system form which can be utilized to realize the global stability of nonlinear systems, and the chapter makes a successful exploration at the actuator fault tolerance. (2) The existing FAS approaches require accurate high-order system states. The chapter proposes a novel nonlinear ESO-based FAS controller, and the strategy enhances the theoretical significance of the HOFAS theory. (3) The chapter presents a novel RAFTC structure with a clear design procedure and a wide range of applications.

Chapter 4
Low-Power FTC for FASs Against Actuator Faults

This chapter introduces an innovative observer-based low-power fault-tolerant control (LPFTC) framework for nonideal time-varying high-order fully actuated systems (FASs). The FAS theory is an emerging control system theory, which can yield global stability of nonlinear systems. In order to enhance the robustness against parameter uncertainties, actuator faults, and sensor noise, an observer-based low-power FAS controller framework is established in the chapter. Additionally, the linear framework adheres to the generalized separation principle and emphasizes the advantages of FAS parametric structures. Furthermore, the work takes into consideration fault tolerance and noise suppression in practical applications, reducing reliance on precise modeling. The uniformly bounded stability and noise suppression performance are also confirmed theoretically and supported experimentally.

4.1 Challenges from Faults and Noises

When practical systems operate in complex and harsh environments, sensor measurement noise is inevitable. For instance, high-frequency tremors may affect sensors when deep-ground drilling systems penetrate hard rocks. Sensor measurement noise is an important factor affecting both observation and control of nonlinear systems, and high-order nonlinear systems are particularly sensitive to measurement noise. Among various types of measurement noise, high-frequency noise can lead to high-power controllers and significant high-frequency oscillations in control input signals. However, actuator energy in practical systems is limited, and high-frequency oscillations of control input signals are undesirable. For example, the servos of a robotic arm cannot be commanded to twist rapidly.

Chapter 2 investigates the problems of observer and controller design for high-order FASs under noise-free conditions. The standard fault-tolerant control (FTC) structure presented therein offers no noise suppression. The existing research on

© The Author(s) 2026
D. Zhou, M. Cai, *Fault-Tolerant Control for Fully Actuated Systems*,
https://doi.org/10.1007/978-981-95-0691-0_4

tracking control for high-order FASs, e.g., Chap. 3, is also unable to effectively handle noise. The bounded stability of its error system is consistent with that of the standard structure discussed in Chap. 2. Therefore, actuator fault compensation and measurement noise suppression represent pressing research areas within the FAS theory.

This chapter proposes an LPFTC framework for uncertain time-varying high-order FASs. In contrast to the standard framework for observation and control of FASs, this LPFTC framework utilizes adjacent error feedback terms. Specifically, the linear LPFTC framework adheres to the generalized separation principle, further highlighting the advantages of parameterized design for FASs. Furthermore, this framework also incorporates both fault compensation and high-frequency noise suppression, thereby reducing the dependence of FAS theory on model accuracy and enhancing its capability to handle parameter uncertainties, actuator faults, and high-frequency noise.

4.2 Problem Formulation and Key Insights

Consider a class of time-varying FAS models:

$$
\begin{cases}
x_o^{(n)} = F_o(x_o^{(0 \sim n-1)}, \xi, t) + G_o(x_o^{(0 \sim n-1)}, \xi, t)\tilde{u} + K_o(x_o^{(0 \sim n-1)}, \xi, t)d, \\
y = H(x_o) + v,
\end{cases}
$$

(4.1)

where $x_o \in \mathbb{R}^m$ is the basic state, $\tilde{u} \in \mathbb{R}^m$ is the control input with possible actuator faults, $d \in \mathbb{R}^m$ is the time-varying process disturbance, $y \in \mathbb{R}^m$ is the measurement output, $\xi \in \mathbb{R}$ is the unmodeled dynamics, and $v = [v_1, v_2, \cdots, v_m]^\mathsf{T} \in \mathbb{R}^m$ is the measurement noise. $F_o(\cdot) \in \mathbb{R}^m, G_o(\cdot) \in \mathbb{R}^{m \times m}, K_o(\cdot) \in \mathbb{R}^{m \times m}, H_o(\cdot) \in \mathbb{R}^m$ are unknown smooth functions. The nonlinear system (4.1) is controllable and observable [21, 22].

Assumption 4.1 ([4]) The process disturbance d satisfies $\sup_{t \geq 0} \|d(t)\| < \infty$ and $\sup_{t \geq 0} \|\dot{d}(t)\| < \infty$. The measurement noise v has high-frequency characteristics, and its specific form is $v_j(t) = \sum_{k=1}^{\mathfrak{N}} v_{j,k} \sin(\omega_{j,k} t)$ where $v_{j,k} > 0, \omega_{j,k} \gg 1, j \in [m], k \in [\mathfrak{N}]$.

Remark 4.1 If the noise is in a more general form, the subsequent analysis in this chapter can still ensure the bounded convergence of the observation error and the tracking error. Assumption 4.1 is to facilitate the comparison between the proposed LPFTC and the standard FTC. When the measurement noise has obvious high-frequency characteristics, the controller energy of LPFTC is smaller.

When dynamic systems operate in complex environments, the actuators often experience gain attenuation faults and deviation faults. To enrich the classic FAS theory, consider the following actuator fault model:

$$\tilde{u} = (1 - \rho)u + u_f, \tag{4.2}$$

where $\rho \in (0, 1)$ is a bounded and differentiable gain decay fault, u_f is a bounded and differentiable deviation fault, and u is the actual control signal.

Define $x = H(x_o)$, and the FAS model (4.1) can be transformed into

$$\begin{cases} x^{(n)} = F(x^{(0 \sim n-1)}, \xi, t) + G(x^{(0 \sim n-1)}, \xi, t)\tilde{u} + K(x^{(0 \sim n-1)}, \xi, t)d, \\ y = x + v, \end{cases} \tag{4.3}$$

where $x \in \mathbb{R}^m$ is the new state, and $F(\cdot) \in \mathbb{R}^m, G(\cdot) \in \mathbb{R}^{m \times m}, K(\cdot) \in \mathbb{R}^{m \times m}$ are the transformed unknown nonlinear functions with bounded partial derivatives. The conversion requirement between (4.1) and (4.3) is the observability, and the conversion method is the Lie derivative or Lie bracket technique.

Assumption 4.2 The entries of matrix functions $F_j, G_{j,k}, K_{j,k}, j, k \in [m]$ satisfy

$$\begin{cases} \left\| \dfrac{\partial F_j}{\partial x^{(i)}} \right\| < \mathscr{C}_1 + \mathscr{D}_1(\xi), \quad \left\| \dfrac{\partial F_j}{\partial \xi} \right\| + \left\| \dfrac{\partial F_j}{\partial t} \right\| < \mathscr{C}_1 + \mathscr{C}_2 \| x^{(0 \sim n-1)} \|, \\[2mm] \left\| \dfrac{\partial G_{j,k}}{\partial x^{(i)}} \right\| < \mathscr{C}_3 + \mathscr{D}_2(\xi), \quad \left\| \dfrac{\partial G_{j,k}}{\partial \xi} \right\| + \left\| \dfrac{\partial G_{j,k}}{\partial t} \right\| < \mathscr{C}_3 + \mathscr{C}_4 \| x^{(0 \sim n-1)} \|, \\[2mm] \left\| \dfrac{\partial K_{j,k}}{\partial x^{(i)}} \right\| < \mathscr{C}_5 + \mathscr{D}_3(\xi), \quad \left\| \dfrac{\partial K_{j,k}}{\partial \xi} \right\| + \left\| \dfrac{\partial K_{j,k}}{\partial t} \right\| < \mathscr{C}_5 + \mathscr{C}_6 \| x^{(0 \sim n-1)} \|, \end{cases} \tag{4.4}$$

where $i = 0, 1 \cdots, n - 1, \mathscr{C}_1, \mathscr{C}_2, \mathscr{C}_3, \mathscr{C}_4, \mathscr{C}_5, \mathscr{C}_6$ are positive constants, $\xi, \dot{\xi}$ are norm-bounded, and $\mathscr{D}_1(\cdot), \mathscr{D}_2(\cdot), \mathscr{D}_3(\cdot)$ are non-negative continuous functions.

Assumption 4.3 For the input matrix function $G(x^{(0 \sim n-1)}, \xi, t)$, there exists a known invertible matrix $\bar{G}(t), \bar{G}^*(t) = \bar{G}^{-1}(t)$ such that for $j, k, l \in [m]$, the following inequalities hold

$$\begin{cases} \sup_{t \geq 0} \left\| \sum_{k=1}^{m} \sum_{l=1}^{m} G_{j,k} \bar{G}_{k,l}^*(t) \right\| < \infty, \\[3mm] \sup_{t \geq 0} \left\| \sum_{k=1}^{m} \sum_{l=1}^{m} \bar{G}_{j,k} \bar{G}_{k,l}^*(t) \right\| < \infty, \\[3mm] \sup_{t \geq 0} \left\| \dfrac{\mathrm{d}}{\mathrm{d}t} \sum_{k=1}^{m} \sum_{l=1}^{m} G_{j,k} \bar{G}_{k,l}^*(t) \right\| < \infty. \end{cases} \tag{4.5}$$

Remark 4.2 $\bar{G}(t)$ in Assumption 4.3 is a nominal value of unknown functions $G(x^{(0 \sim n-1)}, \xi, t)$. This nominal input matrix is not necessarily precise. Similar nominal matrix assumptions are a prerequisite for the observer and controller design [45]. Many practical systems satisfy Assumption 4.3. For example, the input matrix of a robotic arm is known or has a slight perturbation.

If the actuator is ideal, the nonlinear dynamics is known, and the new state $x^{(0 \sim n-1)}$ is fully measurable, then the ideal controller structure is [23]

$$
u = -G^{-1}(x^{(0 \sim n-1)}, \xi, t) \left[F(x^{(0 \sim n-1)}, \xi, t) + A_{0 \sim n-1} x^{(0 \sim n-1)} - u_a \right].
$$
(4.6)

It follows that

$$
\dot{x}^{(0 \sim n-1)} = \Phi_A x^{(0 \sim n-1)} + B_n \left[u_a + K(x^{(0 \sim n-1)}, \xi, t) d \right],
$$
(4.7)

where u_a is an auxiliary signal for achieving other control tasks,

$$
\Phi_A = \begin{bmatrix} 0 & I_m & 0 & \cdots & 0 \\ 0 & 0 & I_m & \cdots & 0 \\ \vdots & & & \ddots & \vdots \\ 0 & 0 & 0 & \cdots & I_m \\ -A_0 & -A_1 & -A_2 & \cdots & -A_{n-1} \end{bmatrix}, \quad B_n = \begin{bmatrix} 0 \\ 0 \\ \vdots \\ 0 \\ I_m \end{bmatrix}.
$$
(4.8)

The advantage of FAS theory lies in transforming complex nonlinear system control problems into simple linear system control problems. However, in most practical engineering applications, the precise dynamic equation cannot be determined, and the sensor can only provide the low-order state measurement y after noise pollution. Thus, the ideal FAS controller (4.6) cannot be given. In addition, actuator faults can also cause the closed-loop linearization structure to fail. It is necessary to study the FAS fault-tolerant controller that can suppress noise, reject disturbances and compensate faults.

Define $X = [x_1^\mathsf{T}, x_2^\mathsf{T}, \cdots, x_n^\mathsf{T}]^\mathsf{T} \triangleq x^{(0 \sim n-1)}$, and the FAS model can be rewritten as

$$
\begin{cases} \dot{X} = \Phi_0 X + B_n \left(\varpi + \bar{G} u \right), \\ y = x_1 + v, \end{cases}
$$
(4.9)

where ϖ represents all uncertainties, including unknown dynamics, disturbances and actuator faults. It can be written as

$$
\varpi = F(X, \xi, t) + G(X, \xi, t) \tilde{u} - \bar{G} u + K(X, \xi, t) d.
$$
(4.10)

The specific form of the matrix Φ_0 is

$$\Phi_0 = \begin{bmatrix} 0 & I_m & 0 & \cdots & 0 \\ 0 & 0 & I_m & \cdots & 0 \\ \vdots & & & \ddots & \vdots \\ 0 & 0 & 0 & \cdots & I_m \\ 0 & 0 & 0 & \cdots & 0 \end{bmatrix}. \tag{4.11}$$

The control task is to let x track $x_d \in \mathbb{R}^m$, and the reference signal satisfies $\sup_{t \geq 0} \|x_d^{(0 \sim n)}\| < \infty$. The solution of its derivatives refers to the TD technique in Chap. 2, and $X_R = [x_{R1}^\mathsf{T}, x_{R2}^\mathsf{T}, \cdots, x_{Rn}^\mathsf{T}]^\mathsf{T}$ is the TD state. In order to guarantee the parametric structure of the closed-loop FAS, a standard nonlinear observer-based FTC framework is designed as follows:

$$\begin{cases} \dot{\hat{x}}_i = \hat{x}_{i+1} + \epsilon^{n-i}(t) h_i \left(\frac{y - \hat{x}_1}{\epsilon^n(t)} \right), i \in [n-1], \\ \dot{\hat{x}}_n = \hat{\varpi} + h_n \left(\frac{y - \hat{x}_1}{\epsilon^n(t)} \right) + \bar{G}u, \\ \dot{\hat{\varpi}} = \frac{1}{\epsilon(t)} h_{n+1} \left(\frac{y - \hat{x}_1}{\epsilon^n(t)} \right), \\ u = -\bar{G}^{-1} \left[\hat{\varpi} - x_{R(n+1)} + A_{0 \sim n-1}(\hat{X} - X_R) \right], \end{cases} \tag{4.12}$$

where $\hat{X} = [\hat{x}_1^\mathsf{T}, \hat{x}_2^\mathsf{T}, \cdots, \hat{x}_n^\mathsf{T}]^\mathsf{T}$ is state observation and $\hat{\varpi}$ is an estimation of total uncertainties. $h_i(\cdot) \in \mathbb{R}^m, i \in [n+1]$ are nonlinear functions to be designed, which satisfy $\|h_i(r)\| \leq \mathscr{H}_i \|r\|$ with $\mathscr{H}_i > 0$. $\epsilon(t)$ is a time-varying gain, and its form is

$$\dot{\epsilon}(t) = \begin{cases} -\alpha \epsilon(t), \epsilon(t) > \epsilon_0, \\ 0, \epsilon(t) \leq \epsilon_0, \end{cases} \tag{4.13}$$

where $\epsilon(0) = 1$, α is the attenuation parameter, and ϵ_0 is a positive constant less than 1.

Assumption 4.4 Consider a time-varying matrix $\Psi \in \mathbb{R}^{m(n+1) \times m(n+1)}$

$$\Psi = \begin{bmatrix} \psi_1 & I_m & 0 & \cdots & 0 \\ \psi_2 & 0 & I_m & \cdots & 0 \\ \vdots & & & \ddots & \\ \psi_n & 0 & 0 & \cdots & I_m \\ \psi_{n+1} & 0 & 0 & \cdots & 0 \end{bmatrix} \tag{4.14}$$

where $\psi_i = -\text{diag}\left(\dot{h}_{i,1}(\sigma_i), \dot{h}_{i,2}(\sigma_i), \cdots, \dot{h}_{i,m}(\sigma_i) \right), i \in [n+1]$, and $\sigma_i \in \mathbb{R}^m$ is a variable related to time. There exists a positive definite constant matrix $\mathfrak{P}$ such that

$$\mathfrak{P}\Psi + \Psi^\mathsf{T}\mathfrak{P} = -\kappa_2 I_{m(n+1)}, \ \forall t > 0, \tag{4.15}$$

where κ_2 is a positive constant.

Theorem 4.1 *For the FAS* (4.9) *under the integrated framework of observer and controller* (4.12), *if Assumption 4.1–Assumption 4.4 hold, there exist positive constants* $\kappa_1, \kappa_2, \nu_1, \nu_2, \nu_3, \nu_4, \varkappa_1, \varkappa_2, \varkappa_3$ *such that*

1. The observation error and tracking error are ultimately uniformly bounded.
2. The error system converges in the following form

$$
\begin{cases}
\|x_i - \hat{x}_i\| \le \max\left\{ \nu_1 \epsilon_0^{n+1-i} \exp(-\tfrac{\nu_2}{\epsilon_0}t)\|x_i(0) - \hat{x}_i(0)\|, \epsilon_0^{n+2-i}\varkappa_1 \right\}, i \in [n], \\[2mm]
\|\varpi - \hat{\varpi}\| \le \max\left\{ \nu_1 \exp(-\tfrac{\nu_2}{\epsilon_0}t)\|\varpi(0) - \hat{\varpi}(0)\|, \epsilon_0\varkappa_1 \right\}, \\[2mm]
\|x_i - x_{Ri}\| \le \max\left\{ \nu_3 \exp(-\nu_4\kappa_2 t)\|x_i(0) - x_{Ri}(0)\|, \epsilon_0\varkappa_2 \right\}, i \in [n], \\[2mm]
\|x_i - x_d^{(i-1)}\| \le \max\left\{ \nu_3 \exp(-\nu_4\kappa_2 t)\|x_i(0) - x_d^{(i-1)}(0)\|, \epsilon_0\varkappa_3 \right\}, i \in [n].
\end{cases}
\tag{4.16}
$$

Proof The proof process is similar to Theorem 2.1 in Chap. 2.

To facilitate parameter tuning, a standard linear integrated framework is designed as follows:

$$
\begin{cases}
\dot{\hat{x}}_i = \hat{x}_{i+1} + \frac{\mathscr{H}_i}{\epsilon^i(t)}\left(y - \hat{x}_1\right), i \in [n-1], \\[2mm]
\dot{\hat{x}}_n = \hat{\varpi} + \frac{\mathscr{H}_n}{\epsilon^n(t)}\left(y - \hat{x}_1\right) + \bar{G}u, \\[2mm]
\dot{\hat{\varpi}} = \frac{\mathscr{H}_{n+1}}{\epsilon^{n+1}(t)}\left(y - \hat{x}_1\right), \\[2mm]
u = -\bar{G}^{-1}\left[\hat{\varpi} - x_{R(n+1)} + A_{0\sim n-1}(\hat{X} - X_R)\right],
\end{cases}
\tag{4.17}
$$

where $\mathscr{H}_i, i \in [n+1]$ should create a Hurwitz matrix

$$
\Psi_e = \begin{bmatrix}
-\mathscr{H}_1 I_m & I_m & 0 & \cdots & 0 \\
-\mathscr{H}_2 I_m & 0 & I_m & \cdots & 0 \\
\vdots & & & \ddots & \\
-\mathscr{H}_n I_m & 0 & 0 & \cdots & I_m \\
-\mathscr{H}_{n+1} I_m & 0 & 0 & \cdots & 0
\end{bmatrix}.
\tag{4.18}
$$

Theorem 4.2 *For the FAS* (4.9) *under the integrated framework of observer and controller* (4.17), *if Assumptions 4.1–4.3 hold, there exist positive constants* $\kappa_1, \kappa_2, \nu_1, \nu_2, \nu_3, \nu_4, \varkappa_1, \varkappa_2, \varkappa_3$ *such that*

1. The observation error and tracking error are ultimately uniformly bounded.
2. The error system converges in the following form

$$\begin{cases} \|x_i - \hat{x}_i\| \leq \max\left\{ \nu_1 \epsilon_0^{n+1-i} \exp(-\frac{\nu_2}{\epsilon_0}t)\|x_i(0) - \hat{x}_i(0)\|, \epsilon_0^{n+2-i}\varkappa_1 \right\}, i \in [n], \\[2mm] \|\varpi - \hat{\varpi}\| \leq \max\left\{ \nu_1 \exp(-\frac{\nu_2}{\epsilon_0}t)\|\varpi(0) - \hat{\varpi}(0)\|, \epsilon_0\varkappa_1 \right\}, \\[2mm] \|x_i - x_{Ri}\| \leq \max\left\{ \nu_3 \exp(-\nu_4\kappa_2 t)\|x_i(0) - x_{Ri}(0)\|, \epsilon_0\varkappa_2 \right\}, i \in [n], \\[2mm] \|x_i - x_d^{(i-1)}\| \leq \max\left\{ \nu_3 \exp(-\nu_4\kappa_2 t)\|x_i(0) - x_d^{(i-1)}(0)\|, \epsilon_0\varkappa_3 \right\}, i \in [n]. \end{cases}$$
$$(4.19)$$

Proof The linear framework no longer requires Assumption 4.4, because the Hurwitz matrix Ψ_e naturally meets Assumption 4.4. The proof process can refer to Corollary 2.1 in Chap. 2.

One advantage of the FAS theory is that the observation matrix Ψ_e and the feedback control matrix Φ_A can be parameterized and designed according to the generalized separation principle. Unlike the separation principle of linear systems, the additional dimension of Ψ_e compared to Φ_A is used to handle the nonlinear uncertain component ϖ, and the extra dimension corresponds to the nonlinear cancellation principle of the FAS theory.

Remark 4.3 The integrated frameworks (4.12) and (4.17) have the following disadvantages:

1. The gain parameter ϵ_0 can reach the order of $n + 1$. In fact, the gain parameter ϵ_0 is a much smaller constant, i.e., the value of ϵ^{n+1} may be very small. The $n + 1$ order operation will amplify the influence of measurement noise on the error term $y - \hat{x}_1$, i.e., the standard observer is extremely sensitive to measurement noise.
2. The observation is affected by the high gain $1/\epsilon_0^i, i \in [n + 1]$, and the controller based on this high-gain observation value will have higher energy. The actual system actuator cannot provide high-power input signal, so this standard framework is difficult to implement in practical engineering.

Therefore, this chapter aims to study a low-power observer-based fault-tolerant tracking controller to reduce the influence of high-frequency measurement noise, which is specifically defined as follows.

Definition 4.1 (LPFTC) For the same system under the same trajectory tracking task, when the high-frequency measurement noise is consistent, consider the energy index $\int_0^\infty \|u\|^2 dt$, if the energy index of an FTC is lower than that of the standard FTC (4.12) and (4.17), then the FTC is called an LPFTC.

4.3 Low-Power Control System Design

In order to promote the practical application of the FAS approach, taking into account the FTC performance, anti-interference performance and noise suppression performance, this chapter proposes a novel LPFTC framework for the FAS model:

$$
\begin{cases}
\dot{\zeta}_i = \begin{bmatrix} B_2^\mathsf{T}\zeta_i + \epsilon(t)h_i(e_i) \\ B_2^\mathsf{T}\zeta_{i+1} + \hbar_i(e_i) \end{bmatrix}, i \in [n-2], \\[2mm]
\dot{\zeta}_{n-1} = \begin{bmatrix} B_2^\mathsf{T}\zeta_{n-1} + \epsilon(t)h_{n-1}(e_{n-1}) \\ B_2^\mathsf{T}\zeta_n + \bar{G}u' + \hbar_{n-1}(e_{n-1}) \end{bmatrix}, \\[2mm]
\dot{\zeta}_n = \begin{bmatrix} B_2^\mathsf{T}\zeta_n + \bar{G}u' + \hbar_{n,1}(e_n) \\ \hbar_{n,2}(e_n) \end{bmatrix}, \\[2mm]
u' = -\bar{G}^{-1}\left[\hat{\varpi}' - x_{R(n+1)} + A_{0\sim n-1}(\hat{X}' - X_R)\right],
\end{cases}
\tag{4.20}
$$

where $h_i, \hbar_i \in \mathbb{R}^m, i \in [n]$ are the functions to be designed. $\zeta = [\zeta_1^\mathsf{T}, \zeta_2^\mathsf{T}, \cdots, \zeta_n^\mathsf{T}]^\mathsf{T}, \zeta_i \in \mathbb{R}^{2m}, i \in [n]. \ e_1 = (y - C_2\zeta_1)/\epsilon^2(t), e_i = (B_2^\mathsf{T}\zeta_{i-1} - C_2\zeta_i)/\epsilon^2(t), 2 \le i \le n, B_2 = [0_{m\times m}, I_m]^\mathsf{T}, C_2 = [I_m, 0_{m\times m}],$

$$
\hat{X}' = L_1\zeta, L_1 = \text{diag}\left(\underbrace{C_2, C_2, \cdots, C_2}_{n}\right), \hat{\varpi}' = B_2^\mathsf{T}\zeta_n,
\tag{4.21}
$$

where $\hat{X}'$ is the estimation of X. Furthermore, there is redundant information in ζ, and the another estimation of X is

$$
\hat{X}'' = L_2\zeta, L_2 = \left[\text{diag}\left(I_{2m}, \underbrace{B_2^\mathsf{T}, \cdots, B_2^\mathsf{T}}_{n-2}\right), 0_{2m(n-1)\times 2m}\right].
\tag{4.22}
$$

Therefore, this framework presents two observations $\hat{X}' = [\hat{x}_1'^\mathsf{T}, \hat{x}_2'^\mathsf{T}, \cdots, \hat{x}_n'^\mathsf{T}]^\mathsf{T}$ and $\hat{X}'' = [\hat{x}_1''^\mathsf{T}, \hat{x}_2''^\mathsf{T}, \cdots, \hat{x}_n''^\mathsf{T}]^\mathsf{T}$.

Assumption 4.5 Consider a time-varying matrix E

$$
E = \begin{bmatrix}
R_1 & N & 0 & \cdots & \cdots & & \cdots & 0 \\
Q_2 & R_2 & N & \ddots & & & & \vdots \\
0 & \ddots & \ddots & \ddots & \ddots & & & \vdots \\
\vdots & & \ddots & \ddots & \ddots & & \ddots & 0 \\
\vdots & & & & \ddots & Q_{n-1} & R_{n-1} & N \\
0 & \cdots & \cdots & \cdots & & 0 & Q_n & R_n
\end{bmatrix},
\tag{4.23}
$$

where

$$R_i = \begin{bmatrix} -\mathrm{diag}\left(\dot{h}_{i,1}(\sigma_{i1}), \dot{h}_{i,2}(\sigma_{i1}), \cdots, \dot{h}_{i,m}(\sigma_{i1})\right) & I_m \\ -\mathrm{diag}\left(\dot{h}_{i,1}(\sigma_{i2}), \dot{h}_{i,2}(\sigma_{i2}), \cdots, \dot{h}_{i,m}(\sigma_{i2})\right) & \mathbf{0}_{m \times m} \end{bmatrix},$$

$$Q_i = \begin{bmatrix} \mathbf{0} \, \mathrm{diag}\left(\dot{h}_{i,1}(\sigma_{i1}), \dot{h}_{i,2}(\sigma_{i1}), \cdots, \dot{h}_{i,m}(\sigma_{i1})\right) \\ \mathbf{0} \, \mathrm{diag}\left(\dot{h}_{i,1}(\sigma_{i2}), \dot{h}_{i,2}(\sigma_{i2}), \cdots, \dot{h}_{i,m}(\sigma_{i2})\right) \end{bmatrix},$$

$$N = \begin{bmatrix} \mathbf{0} & \mathbf{0} \\ \mathbf{0} & I_m \end{bmatrix},$$

$\sigma_{i1}, \sigma_{i2} \in \mathbb{R}^m, i \in [n]$ is a variable related to time, there exists a positive definite constant matrix $\mathscr{P}$ satisfying

$$\mathscr{P}E + E^{\mathsf{T}}\mathscr{P} = -\kappa_2' I_{2mn}, \forall t > 0, \tag{4.24}$$

with a positive constant κ_2'.

Remark 4.4 In a physical sense, Assumption 4.5 is a sufficient condition for the convergence of observation error in the nonlinear LPFTC framework. When the nonlinear LPFTC is specialized as a linear LPFTC, Assumption 4.5 is specialized as Lemma 4.1.

Theorem 4.3 *For the FAS (4.9) under the LPFTC framework of observer and controller (4.20), if Assumptions 4.1–4.3, and 4.5 hold, and $\epsilon_0, \kappa_1, \kappa_2'$ satisfy (4.59), there exist positive constants $\delta, \iota_1, \iota_2, \iota_3, \iota_4, \Omega_0^*, \Omega_1^*, \Omega_2$ such that*

1. The observation error and tracking error are ultimately uniformly bounded.
2. The error system converges in the following form

$$\begin{cases} \|x_i - \hat{x}_i'\| \leq \max\left\{\iota_1 \epsilon_0^{1-i} \exp(-\frac{\iota_2 \kappa_2'}{\epsilon_0}t)\|x_i(0) - \hat{x}_i'(0)\|, \dfrac{\epsilon_0^{2-i}\Omega_0^*}{\delta\kappa_2'}\right\}, i \in [n] \\[2ex] \|\varpi - \hat{\varpi}'\| \leq \max\left\{\iota_1 \epsilon_0^{-n} \exp(-\frac{\iota_2 \kappa_2'}{\epsilon_0}t)\|\varpi(0) - \hat{\varpi}'(0)\|, \dfrac{\epsilon_0^{1-n}\Omega_0^*}{\delta\kappa_2'}\right\}, \\[2ex] \|x_i - \hat{x}_i''\| \leq \max\left\{\iota_1 \epsilon_0^{-i} \exp(-\frac{\iota_2 \kappa_2'}{\epsilon_0}t)\|x_i(0) - \hat{x}_i''(0)\|, \dfrac{\epsilon_0^{1-i}\Omega_0^*}{\delta\kappa_2'}\right\}, i \in [n], \\[2ex] \|x_i - x_{Ri}\| \leq \left\{\iota_3 \exp(-\iota_4 \kappa_1 t)\|x_i(0) - x_{Ri}(0)\|, \dfrac{\Omega_1^*}{\delta\kappa_1}\right\}, i \in [n], \\[2ex] \|x_i - x_d^{(i-1)}\| \leq \left\{\iota_3 \exp(-\iota_4 \kappa_1 t)\|x_i(0) - x_d^{(i-1)}(0)\|, \Omega_2\right\}, i \in [n]. \end{cases} \tag{4.25}$$

Proof Compared with the constant gain ϵ_0, the time-varying gain has an additional exponential decline period. Within the finite interval of the exponential decline, the Lipschitz nonlinear dynamics does not diverect suddenly. After appropriate analysis, it is only necessary to prove the stability of the error system after the gain is fixed. Define the following error variable

$$\tilde{\xi}_i = \zeta_i - [x_i^\mathsf{T}, x_{i+1}^\mathsf{T}]^\mathsf{T}, i \in [n-1], \tilde{\xi}_n = \zeta_n - [x_n^\mathsf{T}, \varpi^\mathsf{T}]^\mathsf{T}, \tag{4.26}$$

where $\zeta_i = [\zeta_{i1}^\mathsf{T}, \zeta_{i2}^\mathsf{T}]^\mathsf{T}$. The derivatives can be calculated as

$$\begin{aligned}
\dot{\tilde{\xi}}_{11} &= B_2^\mathsf{T} \zeta_1 + \epsilon_0 h_1 \left(\frac{y - C_2 \zeta_1}{\epsilon_0^2} \right) - x_2 \\
&= B_2^\mathsf{T} \tilde{\xi}_1 - \frac{1}{\epsilon_0} \dot{h}_1(\sigma_{11}) C_2 \tilde{\xi}_1 + \frac{1}{\epsilon_0} \dot{h}_1(\sigma_{11}) v,
\end{aligned} \tag{4.27}$$

$$\begin{aligned}
\dot{\tilde{\xi}}_{12} &= B_2^\mathsf{T} \zeta_2 + \hbar_1 \left(\frac{y - C_2 \zeta_1}{\epsilon_0^2} \right) - x_3 \\
&= B_2^\mathsf{T} \tilde{\xi}_2 - \frac{1}{\epsilon_0^2} \dot{h}_1(\sigma_{12}) C_2 \tilde{\xi}_1 + \frac{1}{\epsilon_0^2} \dot{\hbar}_1(\sigma_{12}) v,
\end{aligned} \tag{4.28}$$

where $\dot{h}_1(\sigma_{11}), \dot{h}_1(\sigma_{12}) \in \mathbb{R}^m$ are the differential functions of $\hbar_{1,1}, \hbar_{1,2}$, respectively, under the mean value theorem, and σ_{11}, σ_{12} are two variables related to time.

Similarly, for $2 \leq i \leq n-1$, the derivatives of the corresponding error variables is

$$\begin{aligned}
\dot{\tilde{\xi}}_{i1} &= B_2^\mathsf{T} \zeta_i + \epsilon_0 h_i \left(\frac{B_2^\mathsf{T} \zeta_{i-1} - C_2 \zeta_i}{\epsilon_0^2} \right) - x_{i+1} \\
&= B_2^\mathsf{T} \tilde{\xi}_i + \frac{1}{\epsilon_0} \dot{h}_i(\sigma_{i1}) B_2^\mathsf{T} \tilde{\xi}_{i-1} - \frac{1}{\epsilon_0} \dot{h}_i(\sigma_{i1}) C_2 \tilde{\xi}_i,
\end{aligned} \tag{4.29}$$

$$\begin{aligned}
\dot{\tilde{\xi}}_{i2} &= B_2^\mathsf{T} \zeta_{i+1} + \hbar_i \left(\frac{B_2^\mathsf{T} \zeta_{i-1} - C_2 \zeta_i}{\epsilon_0^2} \right) - x_{i+2} \\
&= B_2^\mathsf{T} \tilde{\xi}_{i+1} + \frac{1}{\epsilon_0^2} \dot{\hbar}_i(\sigma_{i2}) B_2^\mathsf{T} \tilde{\xi}_{i-1} - \frac{1}{\epsilon_0^2} \dot{\hbar}_i(\sigma_{i2}) C_2 \tilde{\xi}_i.
\end{aligned} \tag{4.30}$$

For $i = n$, it turns out that

$$\dot{\tilde{\xi}}_{n1} = B_2^\mathsf{T} \zeta_n + \epsilon_0 h_n \left(\frac{B_2^\mathsf{T} \zeta_{n-1} - C_2 \zeta_n}{\epsilon_0^2} \right) - \varpi$$

$$= B_2^\mathsf{T} \tilde{\zeta}_n + \frac{1}{\epsilon_0} \dot{h}_n(\sigma_{n1}) B_2^\mathsf{T} \tilde{\zeta}_{n-1} - \frac{1}{\epsilon_0} \dot{h}_n(\sigma_{n1}) C_2 \tilde{\zeta}_n, \tag{4.31}$$

$$\dot{\tilde{\zeta}}_{n2} = \hbar_n \left(\frac{B_2^\mathsf{T} \zeta_{n-1} - C_2 \zeta_n}{\epsilon_0^2} \right) - \dot{\varpi}$$

$$= \frac{1}{\epsilon_0^2} \dot{\hbar}_n(\sigma_{n2}) B_2^\mathsf{T} \tilde{\zeta}_{n-1} - \frac{1}{\epsilon_0^2} \dot{\hbar}_n(\sigma_{n2}) C_2 \tilde{\zeta}_n - \dot{\varpi}. \tag{4.32}$$

Then, the dynamic equations of $\tilde{\zeta} = [\tilde{\zeta}_1^\mathsf{T}, \tilde{\zeta}_2^\mathsf{T}, \cdots, \tilde{\zeta}_n^\mathsf{T}]^\mathsf{T}$ are

$$\begin{cases} \dot{\tilde{\zeta}}_1 = \Pi_1 \tilde{\zeta}_1 + N \tilde{\zeta}_2 + \Lambda \mathscr{A}_1 v, \\ \dot{\tilde{\zeta}}_i = \Pi_i \tilde{\zeta}_i + N \tilde{\zeta}_{i+1} + \Lambda \mathscr{A}_i B_2^\mathsf{T} \tilde{\zeta}_{i-1}, 2 \le i \le n-1, \\ \dot{\tilde{\zeta}}_n = \Pi_n \tilde{\zeta}_n + \Lambda \mathscr{A}_n B_2^\mathsf{T} \tilde{\zeta}_{n-1} - B_2 \dot{\varpi}, \end{cases} \tag{4.33}$$

where $\Pi_i = M - \Lambda \mathscr{A}_i C_2, i \in [n]$,

$$M = \begin{bmatrix} \mathbf{0} \ I_m \\ \mathbf{0} \ \mathbf{0} \end{bmatrix}, \Lambda = \begin{bmatrix} \dfrac{1}{\epsilon_0} I_m & \mathbf{0} \\ \mathbf{0} & \dfrac{1}{\epsilon_0^2} I_m \end{bmatrix}, \tag{4.34}$$

$$\mathscr{A}_i = \begin{bmatrix} \operatorname{diag}\left(\dot{h}_{i,1}(\sigma_{i1}), \dot{h}_{i,2}(\sigma_{i1}), \cdots, \dot{h}_{i,m}(\sigma_{i1})\right) \\ \operatorname{diag}\left(\dot{\hbar}_{i,1}(\sigma_{i1}), \dot{\hbar}_{i,2}(\sigma_{i1}), \cdots, \dot{\hbar}_{i,m}(\sigma_{i1})\right) \end{bmatrix}. \tag{4.35}$$

Define the observation error

$$\varepsilon = [\varepsilon_1^\mathsf{T}, \varepsilon_2^\mathsf{T}, \cdots, \varepsilon_n^\mathsf{T}]^\mathsf{T}, \varepsilon_i = \frac{1}{\epsilon_0^{2-i}} \Lambda^{-1} \tilde{\zeta}_i, \tag{4.36}$$

and the derivative of ε_i is

$$\begin{cases} \dot{\varepsilon}_1 = \dfrac{1}{\epsilon_0} \left(R_1 \varepsilon_1 + N \varepsilon_2 + \mathscr{A}_1 v \right), \\ \dot{\varepsilon}_i = \dfrac{1}{\epsilon_0} \left(Q_i \varepsilon_{i-1} + R_i \varepsilon_i + N \varepsilon_{i+1} \right), 2 \le i \le n-1, \\ \dot{\varepsilon}_n = \dfrac{1}{\epsilon_0} (Q_n \varepsilon_{n-1} + R_n \varepsilon_n) - \epsilon^n B_2 \dot{\varpi}. \end{cases} \tag{4.37}$$

Based on (4.37), it can be obtained that

$$\dot{\varepsilon} = \frac{1}{\epsilon_0} E\varepsilon + \frac{1}{\epsilon_0} \mathscr{A} \mathfrak{n}(t) - \epsilon_0^n \mathscr{B} \dot{\varpi}, \tag{4.38}$$

where

$$\mathscr{A} = [\mathscr{A}_1^\mathsf{T}, \mathbf{0}_{m \times 2m(n-1)}]^\mathsf{T}, \mathfrak{B} = [\mathbf{0}_{m \times (2m(n-1)+m)}, I_m]^\mathsf{T}. \tag{4.39}$$

Define the tracking error

$$\eta = [\eta_1^\mathsf{T}, \eta_2^\mathsf{T}, \cdots, \eta_n^\mathsf{T}]^\mathsf{T}, \eta_i = x_i - x_{Ri}, \tag{4.40}$$

and the derivatives of η_i are

$$\begin{cases} \dot{\eta}_i = \eta_{i+1}, i \in [n-1], \\ \dot{\eta}_n = \epsilon_0^{2-n} B_2^\mathsf{T} \Lambda \varepsilon_n - A_{0 \sim n-1} \eta - A_{0 \sim n-1} L_1 S \varepsilon, \end{cases} \tag{4.41}$$

where

$$S = \mathrm{diag}(\epsilon_0, 1, \epsilon_0^{-1}, \cdots, \epsilon_0^{2-n}) \otimes \Lambda. \tag{4.42}$$

Therefore, the overall error dynamics is

$$\begin{cases} \dot{\varepsilon} = \dfrac{1}{\epsilon_0} E \varepsilon + \dfrac{1}{\epsilon_0} \mathscr{A} \mathfrak{n}(t) - \epsilon_0^n \mathfrak{B} \varpi, \\ \dot{\eta} = \Phi_A \eta + B_n (\epsilon_0^{2-n} B_2^\mathsf{T} \Lambda \varepsilon_n - A_{0 \sim n-1} L_1 S \varepsilon). \end{cases} \tag{4.43}$$

It can be seen from the overall error dynamics (4.43) that the observation error and the tracking error are coupled, which is a characteristic of nonlinear systems. The jth element of ϖ is

$$\varpi_j = F_j + \sum_{k=1}^{m} G_{j,k} \left[(1-\rho)u_k' + u_{f,k} \right] - \sum_{k=1}^{m} \bar{G}_{j,k} u_k' + \sum_{k=1}^{m} K_{j,k} d_k. \tag{4.44}$$

Based on the boundedness of the TD state and Assumption 4.2, the following conclusion holds

$$\begin{aligned} \|F_j\| &\leq \Gamma_1 + \Gamma_2(\|X\| + \|\xi\|) \\ &\leq \Gamma_1 + \Gamma_2(\|\eta\| + \|X_R\| + \|\xi\|) \\ &\leq \Gamma_1^* + \Gamma_2\|\eta\|, \end{aligned} \tag{4.45}$$

Similarly, it can be obtained that

$$\|G_{j,k}\| \leq \Gamma_3^* + \Gamma_4\|\eta\|, \|K_{j,k}\| \leq \Gamma_5^* + \Gamma_6\|\eta\|, \tag{4.46}$$

where $\Gamma_1, \Gamma_1^*, \Gamma_2, \Gamma_3^*, \Gamma_4, \Gamma_5^*, \Gamma_6$ are positive constants. There are some equations:

$$G_{j,k}\left[(1-\rho)u'_k + u_{f,k}\right] - \bar{G}_{j,k}u'_k$$

$$= \sum_{l=1}^{m}\left\{(G_{j,k} - \bar{G}_{j,k})\bar{G}^*_{k,l}[x_{R(n+1)} - \hat{\varpi}' - A_{0\sim n-1}(\hat{X}' - X_R)]_l\right\}$$

$$+ G_{j,k}\left\{u_{f,k} + \rho\sum_{l=1}^{m}\bar{G}^*_{k,l}[x_{R(n+1)} - \hat{\varpi}' - A_{0\sim n-1}(\hat{X}' - X_R)]_l\right\},$$

$$\tag{4.47}$$

$$\frac{\mathrm{d}F_j}{\mathrm{d}t} = \sum_{i=1}^{n-1}(\frac{\partial F_j}{\partial x_i})^\mathsf{T}x_{i+1} + (\frac{\partial F_j}{\partial x_n})^\mathsf{T}\varpi + (\frac{\partial F_j}{\partial x_n})^\mathsf{T}\bar{G}u' + \frac{\partial F_j}{\partial \xi}\dot{\xi} + \frac{\partial F_j}{\partial t}$$

$$= \sum_{i=1}^{n-1}(\frac{\partial F_j}{\partial x_i})^\mathsf{T}x_{i+1} + (\frac{\partial F_j}{\partial x_n})^\mathsf{T}\varpi + \frac{\partial F_j}{\partial \xi}\dot{\xi} + \frac{\partial F_j}{\partial t}$$

$$+ (\frac{\partial F_j}{\partial x_n})^\mathsf{T}\left[x_{R(n+1)} - \hat{x}'_{n+1} - A_{0\sim n-1}(\hat{X}' - X_R)\right],$$

$$\tag{4.48}$$

$$\frac{\mathrm{d}}{\mathrm{d}t}\left\{G_{j,k}\left[(1-\rho)u'_k + u_{f,k}\right] - \bar{G}_{j,k}u'_k\right\}$$

$$= \frac{\mathrm{d}}{\mathrm{d}t}\sum_{l=1}^{m}\left\{(G_{j,k} - \bar{G}_{j,k})\bar{G}^*_{k,l}[x_{R(n+1)} - \hat{\varpi}' - A_{0\sim n-1}(\hat{X}' - X_R)]_l\right\}$$

$$+ \frac{\mathrm{d}}{\mathrm{d}t}G_{j,k}\left\{u_{f,k} + \rho\sum_{l=1}^{m}\bar{G}^*_{k,l}[x_{R(n+1)} - \hat{\varpi}' - A_{0\sim n-1}(\hat{X}' - X_R)]_l\right\},$$

$$\tag{4.49}$$

$$\frac{\mathrm{d}}{\mathrm{d}t}K_{j,k}d_k$$

$$= \left\{\sum_{i=1}^{n-1}(\frac{\partial K_{j,k}}{\partial x_i})^\mathsf{T}x_{i+1} + (\frac{\partial K_{j,k}}{\partial x_n})^\mathsf{T}\varpi + (\frac{\partial K_{j,k}}{\partial x_n})^\mathsf{T}\bar{G}u' + \frac{\partial K_{j,k}}{\partial \xi}\dot{\xi} + \frac{\partial K_{j,k}}{\partial t}\right\}d_k$$

$$+ K_{j,k}\dot{d}_k$$

$$= \left\{\sum_{i=1}^{n-1}(\frac{\partial K_{j,k}}{\partial x_i})^\mathsf{T}x_{i+1}\right.$$

$$+(\frac{\partial K_{j,k}}{\partial x_n})^\mathsf{T}\varpi + (\frac{\partial K_{j,k}}{\partial x_n})^\mathsf{T}[x_{R(n+1)} - \hat{\varpi}' - A_{0\sim n-1}(\hat{X}' - X_R)]$$

$$\left.+\frac{\partial K_{j,k}}{\partial \xi}\dot{\xi} + \frac{\partial K_{j,k}}{\partial t}\right\}d_k + K_{j,k}\dot{d}_k.$$

$$\tag{4.50}$$

From (4.44), it follows that

$$\sum_{k=1}^{m}\sum_{l=1}^{m}(1-\rho)G_{j,k}\bar{G}_{k,l}^{*}\hat{\varpi}_{l}'$$

$$= \epsilon_0^{2-n} B_2^{\mathsf{T}} \Lambda \varepsilon_n + F_j + \sum_{k=1}^{m} G_{j,k} u_{f,k} + \sum_{k=1}^{m} K_{j,k} d_k$$

$$+ \sum_{k=1}^{m}\left[(1-\rho)G_{j,k} - \bar{G}_{j,k}\right]\sum_{l=1}^{m}\left\{\bar{G}_{k,l}^{*}[x_{R(n+1)} - A_{0\sim n-1}(\hat{X}' - X_R)]_l\right\}.$$

$$(4.51)$$

Combining (4.44)–(4.51), Assumptions 4.1–4.3, there exist positive constants a_0, a_1, a_2 such that

$$\|\dot{\varpi}\| \leq a_0 + a_1\|\varepsilon\| + a_2\|\eta\|. \tag{4.52}$$

A Lyapunov function is chosen as

$$V_{\varepsilon} = \varepsilon^{\mathsf{T}}\mathscr{P}\varepsilon,\ V_{\eta} = \eta^{\mathsf{T}}P\eta,\ V = V_{\varepsilon} + V_{\eta}, \tag{4.53}$$

where P is a positive definite solution to

$$\Phi_A^{\mathsf{T}} P + P\Phi_A = -\kappa_1 I_{mn}. \tag{4.54}$$

The derivative of V_{ε} is

$$\dot{V}_{\varepsilon} = \frac{1}{\epsilon_0}\varepsilon^{\mathsf{T}}(\mathscr{P}E + E^{\mathsf{T}}\mathscr{P})\varepsilon + \frac{2}{\epsilon_0}\varepsilon^{\mathsf{T}}\mathscr{P}(\mathscr{A}v - \epsilon_0^{n+1}\mathscr{B}\dot{\varpi})$$

$$\leq -\frac{\kappa_2'}{\epsilon_0}\|\varepsilon\|^2 + 2\epsilon_0^n\|\mathscr{P}\|\|\mathscr{B}\|\|\varepsilon\|(a_0 + a_1\|\varepsilon\| + a_2\|\eta\|) + \frac{2}{\epsilon_0}\bar{v}\|\mathscr{P}\|\|\mathscr{A}\|\|\varepsilon\|,$$

$$(4.55)$$

where $\|v\| \leq \bar{v}, \bar{v} > 0$. Meanwhile, the derivative of V_{η} is calculated as follows

$$\dot{V}_{\eta} = \eta^{\mathsf{T}}(P\Phi_A + \Phi_A^{\mathsf{T}}P)\eta + 2\eta^{\mathsf{T}}PB_n(\epsilon_0^{2-n}B_2^{\mathsf{T}}\Lambda\varepsilon_n - A_{0\sim n-1}L_1 S\varepsilon)$$

$$\leq -\kappa_1\|\eta\|^2 + 2\|P\|(\epsilon_0^{-n} + \|A_{0\sim n-1}L_1 S\|)\|\eta\|\|\varepsilon\|. \tag{4.56}$$

Combining (4.55) and (4.56), it can be obtained that

$$\dot{V} < -(\frac{\kappa_2'}{\epsilon_0} - \epsilon_0^n b_1)\|\varepsilon\|^2 + (\epsilon_0^n b_0 + b_2)\|\varepsilon\| + b_3\|\varepsilon\|\|\eta\| - \kappa_1\|\eta\|^2$$

$$= -(\frac{\kappa_2'}{\epsilon_0} - \epsilon_0^n b_1)\|\varepsilon\|^2 + (\epsilon_0^n b_0 + b_2)\|\varepsilon\| + \sqrt{\frac{\kappa_2'}{\epsilon_0}}\|\varepsilon\|\sqrt{\frac{\epsilon_0}{\kappa_2'}}b_3\|\eta\| - \kappa_1\|\eta\|^2$$

$$\le -(\frac{\kappa_2'}{2\epsilon_0} - \epsilon_0^n b_1)\|\varepsilon\|^2 + (\epsilon_0^n b_0 + b_2)\|\varepsilon\| - (\kappa_1 - \frac{\epsilon b_3^2}{2\kappa_2'})\|\eta\|^2, \tag{4.57}$$

where $b_0 = 2a_0\|\mathscr{P}\|\|\mathscr{B}\|$, $b_1 = 2a_1\|\mathscr{P}\|\|\mathscr{B}\|$, $b_2 = 2\bar{v}\|\mathscr{P}\|\|\mathscr{A}\|/\epsilon_0$,

$$b_3 = 2\epsilon_0^n a_2\|\mathscr{P}\|\|\mathscr{B}\| + 2\|P\|(\epsilon_0^{-n} + \|A_{0\sim n-1}L_1 S\|). \tag{4.58}$$

Based on the Lyapunov stability theory, the overall error system is ultimately uniformly bounded, i.e., there exist constants $\Theta_1 > 0, t_\epsilon > 0$ such that $\|[\varepsilon^\mathsf{T}, \eta^\mathsf{T}]^\mathsf{T}\| \le \Theta_1, t > t_\epsilon$. The corresponding parameter design standard is

$$\frac{\kappa_2'}{2\epsilon_0} - \epsilon_0^n b_1 > 0, \kappa_1 - \frac{\epsilon b_3^2}{2\kappa_2'} > 0. \tag{4.59}$$

Substituting the bounded observation error ε and the bounded tracking error η into (4.52), it can be obtained that

$$\|\dot{\varpi}\| \le \Omega_0, t > t_\epsilon, \tag{4.60}$$

where $\Omega_0 > 0$. At this point, the derivative of V_ε satisfies

$$\dot{V}_\varepsilon = \frac{1}{\epsilon_0}\varepsilon^\mathsf{T}(\mathscr{P}E + E^\mathsf{T}\mathscr{P})\varepsilon + \frac{2}{\epsilon_0}\varepsilon^\mathsf{T}\mathscr{P}[\mathscr{A}v - \epsilon_0^{n+1}\mathscr{B}\dot{\varpi}]$$

$$\le -\frac{\kappa_2'}{\epsilon_0}\|\varepsilon\|^2 + 2\epsilon_0^n \Omega_0\|\mathscr{P}\|\|\mathscr{B}\|\|\varepsilon\| + \frac{2}{\epsilon_0}\bar{v}\|\mathscr{P}\|\|\mathscr{A}\|\|\varepsilon\|$$

$$= -\frac{\kappa_2'}{\epsilon_0}\|\varepsilon\|^2 + \Omega_0^*\|\varepsilon\|, \tag{4.61}$$

where $\Omega_0^* > 0$. Since $\lambda_{\min}(\mathscr{P})\|\varepsilon\|^2 \le V_\varepsilon \le \lambda_{\max}(\mathscr{P})\|\varepsilon\|^2$, if

$$\|\varepsilon\| > \frac{\epsilon_0\Omega_0^*}{\delta\kappa_2'}, \tag{4.62}$$

it turns out that

$$\dot{V}_\varepsilon \le -\frac{\kappa_2'(1-\delta)}{\epsilon_0}\|\varepsilon\|^2 \le -\frac{\kappa_2'(1-\delta)}{\epsilon_0\lambda_{\max}(\mathscr{P})}V_\varepsilon, \tag{4.63}$$

where $\delta \in (0, 1)$. The above inequality indicates that the observation error ε converges exponentially to an error bound $\frac{\epsilon_0 \Omega_0^*}{\delta \kappa_2'}$, i.e.,

$$\|\varepsilon\| \leq \max \left\{ \iota_1 \exp(-\frac{\iota_2 \kappa_2'}{\epsilon_0} t) \|\varepsilon(0)\|, \frac{\epsilon_0 \Omega_0^*}{\delta \kappa_2'} \right\}, \tag{4.64}$$

where $\iota_1, \iota_2 > 0$. Further, it can be obtained that the state observation error satisfies

$$\|x_i - \hat{x}_i'\| = \|\zeta_{i,1}\| = \epsilon_0^{1-i} \|\varepsilon_i\|$$

$$\leq \max \left\{ \iota_1 \epsilon_0^{1-i} \exp(-\frac{\iota_2 \kappa_2'}{\epsilon_0} t) \|\varepsilon(0)\|, \frac{\epsilon_0^{2-i} \Omega_0^*}{\delta \kappa_2'} \right\}, i \in [n]. \tag{4.65}$$

The estimation error of the total uncertainty satisfies

$$\|\varpi - \hat{\varpi}'\| = \|\zeta_{n,2}\| = \epsilon_0^{-n} \|\varepsilon_n\|$$

$$\leq \max \left\{ \iota_1 \epsilon_0^{-n} \exp(-\frac{\iota_2 \kappa_2'}{\epsilon_0} t) \|\varepsilon(0)\|, \frac{\epsilon_0^{1-n} \Omega_0^*}{\delta \kappa_2'} \right\}, \tag{4.66}$$

The redundant state observation error satisfies

$$\|x_i - \hat{x}_i''\| = \|\zeta_{i,2}\| = \epsilon_0^{-i} \|\varepsilon_i\|$$

$$\leq \max \left\{ \iota_1 \epsilon_0^{-i} \exp(-\frac{\iota_2 \kappa_2'}{\epsilon_0} t) \|\varepsilon(0)\|, \frac{\epsilon_0^{1-i} \Omega_0^*}{\delta \kappa_2'} \right\}, i \in [n]. \tag{4.67}$$

Based on (4.56) and the boundedness of ε, the derivative of V_η is

$$\dot{V}_\eta < -\kappa_1 \|\eta\|^2 + \Omega_1 \|\eta\| \|\varepsilon\| \leq -\kappa_1 \|\eta\|^2 + \Omega_1^* \|\eta\|, \tag{4.68}$$

where $\Omega_1, \Omega_1^* > 0$. Similarly, it turns out that

$$\|x_i - x_{Ri}\| \leq \max \left\{ \iota_3 \exp(-\iota_4 \kappa_1 t) \|\eta(0)\|, \frac{\Omega_1^*}{\delta \kappa_1} \right\}, i \in [n], \tag{4.69}$$

where $\iota_3, \iota_4 > 0$. Combined with the convergence of TD technology, the following conclusion holds

$$\|x_1 - x_d\| \leq \max \left\{ \iota_3 \exp(-\iota_4 \kappa_1 t) \|\eta(0)\|, \Omega_2 \right\}, \tag{4.70}$$

where $\Omega_2 > 0$. Theorem 4.3 has been proved.

4.4 Noise Sensitivity and Linear Framework Analysis

In order to facilitate parameter design, a linear LPFTC framework for FASs is designed as

$$
\begin{cases}
\dot{\zeta}_i = \begin{bmatrix} B_2^{\mathsf{T}} \zeta_i + \dfrac{\mathscr{H}'_{i,1}}{\epsilon(t)} e_i \\[2mm] B_2^{\mathsf{T}} \zeta_{i+1} + \dfrac{\mathscr{H}'_{i,2}}{\epsilon^2(t)} e_i \end{bmatrix}, i \in [n-2], \\[8mm]
\dot{\zeta}_{n-1} = \begin{bmatrix} B_2^{\mathsf{T}} \zeta_{n-1} + \dfrac{\mathscr{H}'_{n-1,1}}{\epsilon(t)} e_{n-1} \\[2mm] B_2^{\mathsf{T}} \zeta_n + \bar{G} u' + \dfrac{\mathscr{H}'_{n-1,2}}{\epsilon^2(t)} e_{n-1} \end{bmatrix}, \\[8mm]
\dot{\zeta}_n = \begin{bmatrix} B_2^{\mathsf{T}} \zeta_n + \bar{G} u' + \dfrac{\mathscr{H}'_{n,1}}{\epsilon(t)} e_n \\[2mm] \dfrac{\mathscr{H}'_{n,2}}{\epsilon^2(t)} e_n \end{bmatrix}, \\[8mm]
u' = -\bar{G}^{-1}\left[\hat{x}'_{n+1} - x_{R(n+1)} + A_{0\sim n-1}(\hat{X}' - X_R) \right],
\end{cases}
\tag{4.71}
$$

where $e_1 = y - C_2 \zeta_1$, $e_i = B_2^{\mathsf{T}} \zeta_{i-1} - C_2 \zeta_i$, $2 \leq i \leq n$, and the two state observations are consistent with (4.21) and (4.22). The parameters $\mathscr{H}'_{i,1}, \mathscr{H}'_{i,2}, i \in [n]$ need to make a constant matrix E_e

$$
E_e = \begin{bmatrix}
R_{e1} & N & 0 & \cdots & & \cdots & & 0 \\
Q_{e2} & R_{e2} & N & \ddots & & & & \vdots \\
0 & \ddots & \ddots & \ddots & & \ddots & & \vdots \\
\vdots & & \ddots & \ddots & & \ddots & & 0 \\
\vdots & & & \ddots & & Q_{e(n-1)} & R_{e(n-1)} & N \\
0 & & \cdots & \cdots & & 0 & Q_{en} & R_{en}
\end{bmatrix},
\tag{4.72}
$$

where

$$
R_{ei} = \begin{bmatrix} -\mathscr{H}'_{i,1} I_m & I_m \\ -\mathscr{H}'_{i,2} I_m & 0 \end{bmatrix}, \quad Q_{ei} = \begin{bmatrix} 0 & \mathscr{H}'_{i,1} I_m \\ 0 & \mathscr{H}'_{i,2} I_m \end{bmatrix}.
\tag{4.73}
$$

Lemma 4.1 ([4]) *For the constant matrix E_e, there exists a positive definite constant matrix $\mathscr{P}$ satisfying*

$$
\mathscr{P} E_e + E_e^{\mathsf{T}} \mathscr{P} = -\kappa'_2 I_{2mn},
\tag{4.74}
$$

where κ'_2 is a positive constant.

Theorem 4.4 *For the FAS* (4.9) *under the linear LPFTC framework* (4.71), *if Assumptions 4.1–4.3 hold, and $\epsilon_0, \kappa_1, \kappa_2'$ satisfy* (4.59), *there exist positive constants $\delta, \iota_1, \iota_2, \iota_3, \iota_4, \Omega_0^*, \Omega_1^*, \Omega_2$ such that*

1. *The observation error and tracking error are ultimately uniformly bounded.*
2. *The error system converges in the following form*

$$
\begin{cases}
\|x_i - \hat{x}_i'\| \leq \max\left\{\iota_1 \epsilon_0^{1-i} \exp(-\frac{\iota_2 \kappa_2'}{\epsilon_0} t)\|x_i(0) - \hat{x}_i'(0)\|, \frac{\epsilon_0^{2-i}\Omega_0^*}{\delta\kappa_2'}\right\}, i \in [n] \\[2mm]
\|\varpi - \hat{\varpi}'\| \leq \max\left\{\iota_1 \epsilon_0^{-n} \exp(-\frac{\iota_2 \kappa_2'}{\epsilon_0} t)\|\varpi(0) - \hat{\varpi}'(0)\|, \frac{\epsilon_0^{1-n}\Omega_0^*}{\delta\kappa_2'}\right\}, \\[2mm]
\|x_i - \hat{x}_i''\| \leq \max\left\{\iota_1 \epsilon_0^{-i} \exp(-\frac{\iota_2 \kappa_2'}{\epsilon_0} t)\|x_i(0) - \hat{x}_i''(0)\|, \frac{\epsilon_0^{1-i}\Omega_0^*}{\delta\kappa_2'}\right\}, i \in [n], \\[2mm]
\|x_i - x_{Ri}\| \leq \left\{\iota_3 \exp(-\iota_4 \kappa_1 t)\|x_i(0) - x_{Ri}(0)\|, \frac{\Omega_1^*}{\delta\kappa_1}\right\}, i \in [n], \\[2mm]
\|x_i - x_d^{(i-1)}\| \leq \left\{\iota_3 \exp(-\iota_4 \kappa_1 t)\|x_i(0) - x_d^{(i-1)}(0)\|, \Omega_2\right\}, i \in [n].
\end{cases}
\tag{4.75}
$$

Proof Compared with Theorems 4.3 and 4.4 does not require Assumption 4.5. Because Lemma 4.1 guarantees that there must exist parameters $\mathcal{H}_{i,1}', \mathcal{H}_{i,2}', i \in [n]$ satisfying Assumption 4.5. The proof process of the remaining main body is similar to Theorem 4.3. Furthermore, the two matrices E_e, Φ of this linear LPFTC framework also follow the principle of generalized separability.

High-order system models and methods are highly sensitive to measurement noise. To further prove that the proposed LPFTC framework is of low power and verify the noise suppression performance of the LPFTC framework, In this section, the proposed linear LPFTC structure (4.71) is compared with the standard linear FTC structure (4.17).

Theorem 4.5 *For the closed-loop FAS* (4.9), *if the gain ϵ_0 of the two observers* (4.17) *and* (4.71) *are consistent, and the corresponding matrices* (4.18) *and* (4.72) *are Hurwitz, there exist $\omega^\star > 0$ and $\bar{c}_{i,j,k} > 0, i \in [n], j \in [m], k \in [\mathfrak{N}]$ such that*

$$
\frac{\limsup\limits_{t\to\infty} \|\hat{x}_i'(t) - x_i(t)\|}{\limsup\limits_{t\to\infty} \|\hat{x}_i(t) - x_i(t)\|} \leq \sum_{j=1}^{m}\sum_{k=1}^{\mathfrak{N}} \bar{c}_{i,j,k}\omega_{j,k}^{-(r_i'-1)}, \forall \omega_{j,k} > \omega^\star,
\tag{4.76}
$$

where $r_i' = \min\{i, n, 2n-i+1\}$. Moreover, the controller energy of LPFTC is lower than that of the standard FTC.

Proof Define the observation error of the standard structure (4.17) with constant gain

$$
\aleph = \frac{1}{\epsilon_0} \Lambda_n^{-1}(\hat{X} - X),
\tag{4.77}
$$

where $\Lambda_n = \text{diag}(\frac{1}{\epsilon_0}I_m, \frac{1}{\epsilon_0^2}I_m, \cdots, \frac{1}{\epsilon_0^n}I_m)$. Then, it turns out that

$$\dot{\aleph} = \frac{1}{\epsilon_0}[\Phi_0 + \mathcal{H}_{oc}C_n]\aleph + \frac{1}{\epsilon_0}\mathcal{H}_{oc}v, \qquad (4.78)$$

where $\mathcal{H}_{oc} = [\mathcal{H}_1 I_m, \mathcal{H}_2 I_m, \cdots, \mathcal{H}_n I_m]^\mathsf{T}$. Similarly,

$$\dot{\varepsilon} = \frac{1}{\epsilon_0}E_e\varepsilon + \frac{1}{\epsilon_0}\mathscr{A}v - \epsilon_0^n\mathfrak{B}\dot{\omega}. \qquad (4.79)$$

When the noise v is taken as the input signal and the observation error as the output signal, the observation error of the standard observer is

$$\hat{x}_i - x_i = \frac{1}{\epsilon_0^{i-1}}\aleph_i, i \in [n]. \qquad (4.80)$$

The corresponding relative order is $r_i = 1, i \in [n]$. Due to the Hurwitz property of (4.78), there exist some harmonic transfer functions $\mathscr{F}_{i,j,k}, j \in [m]$ satisfying

$$\limsup_{t\to\infty} \|\hat{x}_{i,j}(t) - x_{i,j}(t)\| = \sum_{k=1}^{\mathfrak{N}} v_{j,k}\|\mathscr{F}_{i,j,k}(\omega_{j,k})\|. \qquad (4.81)$$

On the other hand, the observation error of the new structure is

$$\hat{x}_i' - x_i = \frac{1}{\epsilon_0}\varepsilon_{i,1}, i \in [n]. \qquad (4.82)$$

The corresponding relative order is $r_i' = \min\{i, n, 2n - i + 1\}, i \in [n]$. Due to the Hurwitz property of (4.79), there exist some harmonic transfer functions $\mathscr{F}_{i,j,k}', j \in [m]$ satisfying

$$\limsup_{t\to\infty} \|\hat{x}_{i,j}'(t) - x_{i,j}(t)\| = \sum_{k=1}^{\mathfrak{N}} v_{j,k}|\mathscr{F}_{i,j,k}'(\omega_{j,k})|. \qquad (4.83)$$

Combining (4.81) and (4.83), there exist $\omega^\star > 0, c_{i,j,k} > 0, c_{i,j,k}' > 0$ such that $\forall \omega_{j,k} > \omega^\star$,

$$\begin{cases} \|F_{i,j,k}(\omega_{j,k})\| \geq c_{i,j,k}\omega_{j,k}^{-1}, \\ \|F_{i,j,k}'(\omega_{j,k})\| \leq c_{i,j,k}'\omega_{j,k}^{-r_i'}. \end{cases} \qquad (4.84)$$

Therefore, it can be obtained that

$$\frac{\limsup_{t\to\infty} |\hat{x}'_{i,j}(t) - x_{i,j}(t)|}{\limsup_{t\to\infty} |\hat{x}_{i,j}(t) - x_{i,j}(t)|} \leq \frac{\sum_{k=1}^{\mathfrak{N}} c'_{i,j,k}\omega_{j,k}^{-r'_i}}{\sum_{k=1}^{\mathfrak{N}} c_{i,j,k}\omega_{j,k}^{-1}} \leq \sum_{k=1}^{\mathfrak{N}} \frac{c'_{i,j,k}}{c_{i,j,k}}\omega_{j,k}^{-(r'_i-1)}. \tag{4.85}$$

This means that the novel observer is less sensitive to high-frequency noise than the standard observer, and the structure of the two controllers is

$$\begin{cases} u = -\bar{G}^{-1}\left[A_{0\sim n-1} \ I_m\right]\begin{bmatrix} \hat{X} - X_R \\ \hat{\varpi} - x_{R(n+1)} \end{bmatrix}, \\ u' = -\bar{G}^{-1}\left[A_{0\sim n-1} \ I_m\right]\begin{bmatrix} \hat{X}' - X_R \\ \hat{\varpi}' - x_{R(n+1)} \end{bmatrix}, \end{cases} \tag{4.86}$$

where

$$\begin{cases} \hat{X} - X_R = \hat{X} - X + X - X_R, \\ \hat{X}' - X_R = \hat{X}' - X + X - X_R. \end{cases} \tag{4.87}$$

Based on Theorems 4.2 and 4.4, both controllers can eventually yield bounded stability. According to Definition 4.1, it can be concluded from (4.85)–(4.87) that the FAS controller u' is of low power. Theorem 4.5 has been proved.

Remark 4.5 When high-frequency noise occurs, the integral structure of the standard observer continuously propagates the noise excitation error, while the observer in the LPFTC framework only utilizes the observed values of two adjacent states, which can significantly weaken the propagation phenomenon of noise in the integral observer. Furthermore, since the maximum observation gain of LPFTC is $1/\epsilon_0^2$, while that of standard FTC reaches $1/\epsilon_0^{n+1}$, and both controllers are designed based on observers, the controller energy of LPFTC is lower than that of standard FTC.

4.5 Experimental Validation

Consider the following fourth-order nonideal FAS model:

$$\begin{cases} x^{(4)} = F(x^{(0\sim 3)}, \xi, t) + G(x^{(0\sim 3)}, \xi, t)u + K(x^{(0\sim 3)}, \xi, t)d, \\ y = x + v, \end{cases} \tag{4.88}$$

where

$$
\begin{cases}
F(\cdot) = 2\sin t \sin x^2 + \dfrac{\dot{x}}{1+\dot{x}^2} + \tanh \ddot{x} + \dddot{x} + x\xi, \\[2mm]
G(\cdot) = 5 + 0.1\sin t, \\[2mm]
K(\cdot) = \lg(1+t)\sin x + \cos \dot{x} + \dddot{x} + 2 + \xi,
\end{cases}
\tag{4.89}
$$

the unmodeled dynamics is $\xi = 0.2\sin t$, the process disturbance is $d = \sin t$, the measurement noise is $v = 0.1\sin 1000t + 0.05\sin 2000t$, and the initial state is $X(0) = [1, 0, 0, 0]^{\mathsf{T}}$. The actuator faults are

$$
\rho =
\begin{cases}
0, t < 20\mathrm{s}, \\
0.3 - 0.3\exp(-100t)\sin(100t), t \geq 20\mathrm{s},
\end{cases}
\tag{4.90}
$$

and

$$
u_f =
\begin{cases}
0, t < 25\mathrm{s}, \\
20 - 20\exp(-t), t \geq 25\mathrm{s}.
\end{cases}
\tag{4.91}
$$

The reference signal is $x_d = 5\sin t + \cos t + 2$, and the control task is to let x track x_d when the system parameters are unknown. In order to simplify the solution process of the high-order derivatives of the reference signal, a TD is designed as

$$
\begin{cases}
\dot{x}_{Ri} = x_{R(i+1)}, x_{Ri}(0) = 0.5, i \in [4], \\
\dot{x}_{R5} = -120R^5(x_{R1} - x_d) - 274R^4 x_{R2} - 225R^3 x_{R3} \\
\qquad - 85R^2 x_{R4} - 85R^2 x_{R4} - 15R x_{R5}, x_{R5}(0) = 0.5,
\end{cases}
\tag{4.92}
$$

where $R = 50$. Then, a linear LPFTC framework is designed as

$$
\begin{cases}
\dot{\zeta}_1 = \begin{bmatrix} B_2^{\mathsf{T}}\zeta_1 + \frac{\mathscr{H}'_{1,1}}{\epsilon(t)}e_1 \\[2mm] B_2^{\mathsf{T}}\zeta_2 + \frac{\mathscr{H}'_{1,2}}{\epsilon^2(t)}e_1 \end{bmatrix}, \\[6mm]
\dot{\zeta}_2 = \begin{bmatrix} B_2^{\mathsf{T}}\zeta_2 + \frac{\mathscr{H}'_{2,1}}{\epsilon(t)}e_2 \\[2mm] B_2^{\mathsf{T}}\zeta_3 + \frac{\mathscr{H}'_{2,2}}{\epsilon^2(t)}e_2 \end{bmatrix}, \\[6mm]
\dot{\zeta}_3 = \begin{bmatrix} B_2^{\mathsf{T}}\zeta_3 + \frac{\mathscr{H}'_{3,1}}{\epsilon(t)}e_3 \\[2mm] B_2^{\mathsf{T}}\zeta_4 + \bar{G}u' + \frac{\mathscr{H}'_{3,2}}{\epsilon^2(t)}e_3 \end{bmatrix}, \\[6mm]
\dot{\zeta}_4 = \begin{bmatrix} B_2^{\mathsf{T}}\zeta_4 + \bar{G}u' + \frac{\mathscr{H}'_{4,1}}{\epsilon(t)}e_4 \\[2mm] \frac{\mathscr{H}'_{4,2}}{\epsilon^2(t)}e_4 \end{bmatrix}, \\[6mm]
u' = -\bar{G}^{-1}\left[\hat{\varpi}' - x_{R5} + A_{0\sim3}(\hat{X}' - X_R)\right],
\end{cases}
\tag{4.93}
$$

where the initial value is zero, and the other parameters are designed as $\mathfrak{a} = 5$, $\epsilon_0 = 0.005$, $A_{0\sim3} = [4, 7, 6, 2]$, $\bar{G}(t) = 5$, $\mathscr{H}'_{1,1} = 0.6251$, $\mathscr{H}'_{1,2} = 0.4553$, $\mathscr{H}'_{2,1} = 0.4395$, $\mathscr{H}'_{2,2} = 0.0848$, $\mathscr{H}'_{3,1} = 0.5144$, $\mathscr{H}'_{3,2} = 0.0439$, $\mathscr{H}'_{4,1} = 0.6$, $\mathscr{H}'_{4,2} = 0.017$.

In order to compare the proposed LPFTC framework with the standard FTC framework, a standard FTC framework is designed as

$$\begin{cases} \dot{\hat{x}}_i = \hat{x}_{i+1} + \epsilon^{-i}(t)\mathscr{H}_i\left(y - \hat{x}_1\right), i \in [3], \\ \dot{\hat{x}}_4 = \hat{\varpi} + \epsilon^{-4}(t)\mathscr{H}_4\left(y - \hat{x}_1\right) + \bar{G}u, \\ \dot{\hat{\varpi}} = \epsilon^{-5}(t)\mathscr{H}_5\left(y - \hat{x}_1\right), \\ u = -\bar{G}^{-1}\left[\hat{\varpi} - x_{R5} + A_{0\sim3}(\hat{X} - X_R)\right], \end{cases} \tag{4.94}$$

where the initial value is zero, and the other parameters are designed as $\mathfrak{a} = 5$, $\epsilon_0 = 0.005$, $A_{0\sim3} = [4, 7, 6, 2]$, $\bar{G}(t) = 5$, $\mathscr{H}_1 = 1.5$, $\mathscr{H}_2 = 0.85$, $\mathscr{H}_3 = 0.225$, $\mathscr{H}_4 = 0.0274$, $\mathscr{H}_5 = 0.0012$.

In practical engineering applications, the output range of the actuator signal is limited. The upper bound of the actuator signal, in a sense, reflects the upper bound of the controller energy. Therefore, the experiment utilized the saturation function $\text{Sat}_{\mathscr{M}}(\cdot) \triangleq \min\{\max\{\cdot, -\mathscr{M}\}, \mathscr{M}\}$ to limit the amplitudes of the above two controllers. The two specific forms are

$$u' = -\bar{G}^{-1}\text{Sat}_{200}\left(\hat{\varpi}' - x_{R5} + A_{0\sim3}(\hat{X}' - X_R)\right), \tag{4.95}$$

$$u = -\bar{G}^{-1}\text{Sat}_{80000}\left(\hat{\varpi} - x_{R5} + A_{0\sim3}(\hat{X} - X_R)\right). \tag{4.96}$$

The saturation bound is related to the noise amplitude and the noise frequency. Under normal circumstances, prior mechanism knowledge, practical engineering experience and computer simulation can determine the specific value of the saturation bound. The reason why the saturation upper bound of $u'$is much lower than that of u is that the controller energy of the standard FTC is much higher than that of the LPFTC. In many simulation tests, once the standard controller is restricted to a smaller range, the FAS (4.88) will diverge. Since the standard FTC is highly sensitive to high-frequency noise, the controller requires a wider range and more refined values to stabilize the high-frequency oscillating system. It can be known from Table 4.1 that the ultimate tracking error of LPFTC is slightly larger than that of standard FTC. The reason is that the variation range of the controller signal of LPFTC is much smaller, and its parameter refinement degree is not as

Table 4.1 The ultimate tracking error

Framework	$\mathscr{M} = 200$	$\mathscr{M} = 80000$
Standard FTC	Divergence	0.4385
LPFTC	0.5754	0.5757

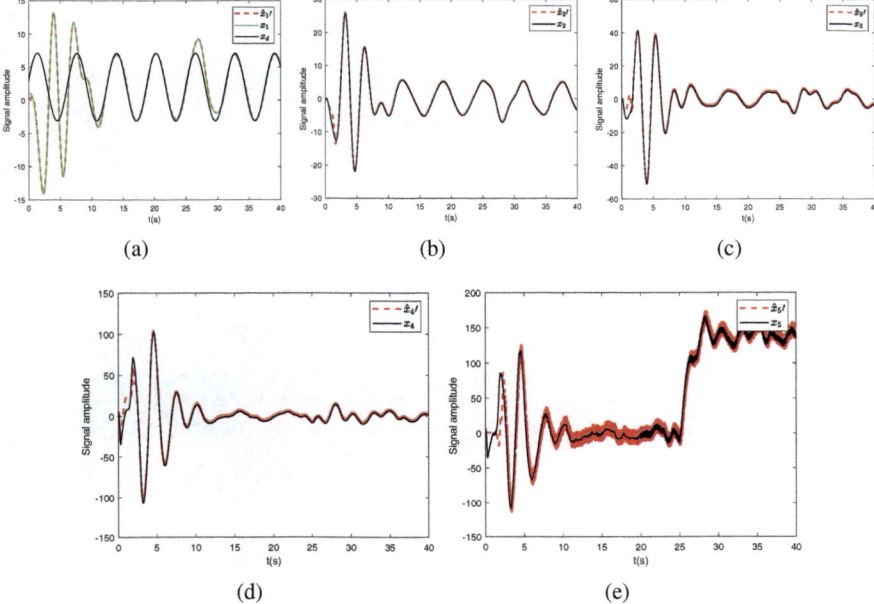

Fig. 4.1 The signal trajectories under the LPFTC framework

good as that of standard FTC. However, more importantly, the controller energy of LPFTC is much smaller than that of standard FTC, and its engineering application value is significantly greater than that of standard FTC.

As shown in Figs. 4.1 and 4.2, the LPFTC framework shows better control performance and noise suppression performance on the uncertain FAS with actuator faults. It explains Theorems 4.1–4.5. It can be seen from Fig. 4.3 that the energy of the proposed controller u' is much smaller than that of the standard controller $\hat{u}$. Therefore, the LPFTC framework proposed in this chapter is more suitable for practical engineering.

4.6   Notes and References

This chapter proposes a novel LPFTC framework for uncertain high-order FAS with actuator faults and high-frequency noise. Compared with the standard FTC framework, the novel framework is less sensitive to high-frequency noise, and the corresponding FAS controller has lower power. This LPFTC strategy is not only a theoretical supplement to the FTC theory, but also an exploration of the engineering application of the FAS approaches.

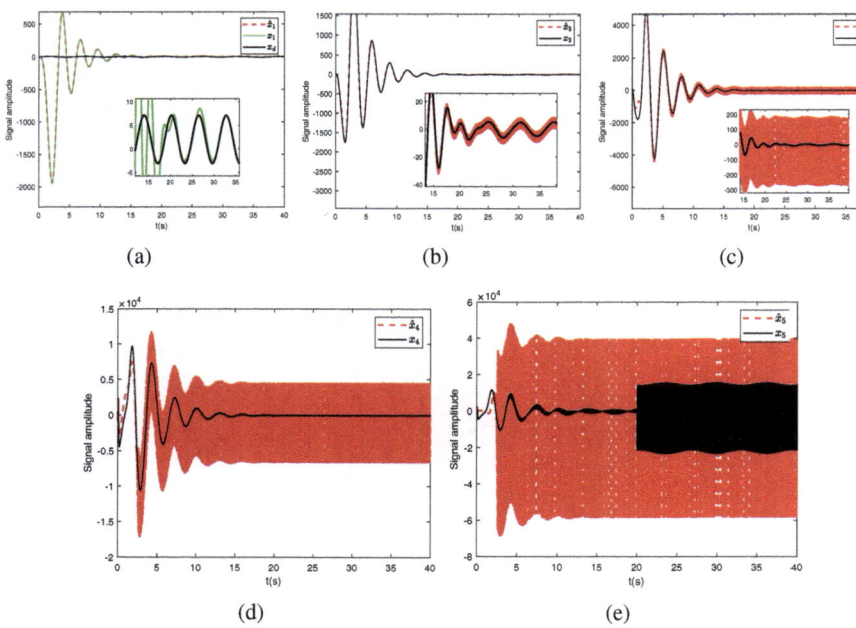

Fig. 4.2 The signal trajectories under the standard FTC framework

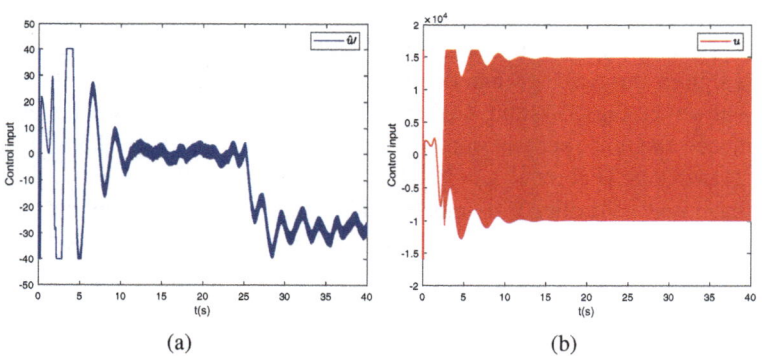

Fig. 4.3 The signal trajectories of two controllers

Chapter 5
Finite-Time FTC for FASs Against Actuator Faults

This chapter develops a finite-time fault-tolerant control (FTFTC) framework to realize system stabilization and trajectory tracking. Paralleling to state-space theory, the fully actuated system (FAS) theory contains rich controller design methods. Most existing FAS approaches can only yield general asymptotic stability. In order to improve the feasibility of FAS approaches in rapid control systems, a parametric FAS stabilization controller based on the homogeneity principle is presented for finite-time stability. Further, a finite-time observer-based FAS tracking controller is structured for faulty FASs. The finite-time observer can yield zero-value convergence of state estimation error and fault estimation error, and the proposed FTFTC framework can give zero-value convergence of the error system in a finite time. The main results are proved theoretically and illustrated experimentally.

5.1 Key Concepts of Finite-Time Control

Most existing parametric FAS controllers aim to achieve asymptotic stability of nonlinear systems. In some rapid control systems, such as high-speed trains and permanent magnet synchronous motors, the general asymptotic stability cannot meet their high-performance requirements. Finite-time stability, as a more stringent system stability criterion than asymptotic stability, has broader application prospects in rapid nonlinear systems [18, 34, 136]. In most cases, the response speed of a finite-time controller is significantly faster than that of an asymptotic controller [117, 122]. The key to the early FAS theory was to design a parametric controller with a fixed structure to enable the nonlinear system to asymptotically converge. The finite-time FAS controller is bound to change the existing parametric structure, and this structural change should not only retain the advantages of the original parametric design, but also ensure the finite-time convergence.

© The Author(s) 2026
D. Zhou, M. Cai, *Fault-Tolerant Control for Fully Actuated Systems*,
https://doi.org/10.1007/978-981-95-0691-0_5

The finite-time stability problem has been well studied in the first-order state-space theory, i.e., the finite-time event-triggered control strategy based on the backstepping method [60], the finite-time tracking control via neural network technology [117], and the finite-time prescribed performance control through fuzzy logic systems [91], etc. Although these state-space methods can effectively achieve finite-time stability, their controller structures are complex. The FAS controller has certain advantages in structural design and parameter tuning. Therefore, it is meaningful to solve an FAS controller structure that guarantees finite-time stability.

This chapter proposes an FTFTC framework based on the FAS theory. The FAS approaches in the previous chapters can only provide general results of global asymptotic stability. In order to improve the applicability of the FAS approaches in rapid control systems, a parameterized finite-time observer-based controller for FASs based on the principle of homogeneity is established. The observer can achieve the finite-time convergence of state estimation error and fault estimation error, and the fault-tolerant controller can achieve the finite-time convergence of tracking error.

5.2 Formulation of Finite-Time Control

Consider a class of FAS models

$$\begin{cases} x^{(n)} = F(x^{(0\sim n-1)}) + G(x^{(0\sim n-1)})\tilde{u}, \\ y = x, \end{cases} \tag{5.1}$$

where $x^{(0\sim n-1)} \in \mathbb{R}^n$ is the full-order system state, $\tilde{u} \in \mathbb{R}$ is the control input with possible actuator faults, $y \in \mathbb{R}$ is the measurement out, and $F(x^{(0\sim n-1)}), G(x^{(0\sim n-1)})$ are nonlinear functions. The actuator fault can be modeled as

$$\tilde{u} = (1 - \rho)u + u_f, \tag{5.2}$$

where $\rho \in (0, 1)$ is a bounded and differentiable gain decay fault, u_f is a bounded and differentiable deviation fault, and u is the actual control input.

Define $\delta = -\rho G(x^{(0\sim n-1)})u + G(x^{(0\sim n-1)})u_f$, and δ represents the overall actuator fault signal satisfying $\sup_{t \geq 0} \|[\delta(t), \dot{\delta}(t)]\| < \bar{\delta}, \bar{\delta} > 0$.

Assumption 5.1 For any $\mathfrak{x}^{(0\sim n-1)}, \mathfrak{y}^{(0\sim n-1)} \in \mathbb{R}^n$, there exist $\bar{F}, \bar{G}, \varkappa > 0$ such that

$$\begin{cases} \|F(\mathfrak{x}^{(0\sim n-1)}) - F(\mathfrak{y}^{(0\sim n-1)})\| \leq \bar{F}\|\mathfrak{x}^{(0\sim n-1)} - \mathfrak{y}^{(0\sim n-1)}\|, \\ \|G(\mathfrak{x}^{(0\sim n-1)}) - G(\mathfrak{y}^{(0\sim n-1)})\| \leq \bar{G}\|\mathfrak{x}^{(0\sim n-1)} - \mathfrak{y}^{(0\sim n-1)}\|, \\ \|G(\mathfrak{x}^{(0\sim n-1)})\| > \varkappa. \end{cases} \tag{5.3}$$

Remark 5.1 Assumption 5.1 contains the Lipschitz nonlinear condition and the full-actuation condition commonly in nonlinear observation and control [94, 136]. Many actual physical objects, i.e., robotic arms, RLC circuits, etc., all meet the above conditions.

If the system (5.1) has no faults and the full-order state $x^{(0 \sim n-1)}$ is measurable, then the traditional FAS controller under the asymptotic stabilization task is [23]

$$u_{as} = -G^{-1}(x^{(0 \sim n-1)})(F(x^{(0 \sim n-1)}) + A_{0 \sim n-1}x^{(0 \sim n-1)}). \tag{5.4}$$

where

$$\Phi_A = \begin{bmatrix} 0 & 1 & 0 & \cdots & 0 \\ 0 & 0 & 1 & \cdots & 0 \\ \vdots & & & \ddots & \vdots \\ 0 & 0 & 0 & \cdots & 1 \\ -A_0 & -A_1 & -A_2 & \cdots & -A_{n-1} \end{bmatrix}, \tag{5.5}$$

and the FAS model can be written as

$$\dot{x}^{(0 \sim n-1)} = \Phi_A x^{(0 \sim n-1)}. \tag{5.6}$$

Based on the FAS theory, there exists a positive definite matrix $P \in \mathbb{R}^{n \times n}$ that is a solution to the following Lyapunov equation

$$P\Phi_A + \Phi_A^{\mathsf{T}} P = -I_n. \tag{5.7}$$

However, in some rapid control systems, such as high-speed trains and spacecraft, general asymptotic stability controllers are difficult to guarantee ideal performance. Finite-time stability has higher requirements than asymptotic stability. Therefore, it is necessary to establish the finite-time FAS controller structure.

Definition 5.1 (Finite-Time Stability [9]) If a nonlinear system

$$\dot{z} = f(z), z \in \mathbb{R}^n \tag{5.8}$$

is Lyapunov stable, and there exists a settling time function $\mathfrak{T}(z(0)) > 0$ such that for $\forall z(0) \in \mathbb{R}^n$,

$$z(t) = 0, \forall t \geq \mathfrak{T}(z(0)), \tag{5.9}$$

then the system (5.8) is globally finite-time stable.

Lemma 5.1 (Finite-Time Stability Criterion [10]) *The nonlinear system (5.8) is globally finite-time stable if and only if there exist a radially unbounded Lyapunov* *function* $\mathfrak{V}(z) : \mathbb{R}^n \to \mathbb{R}$ *and two constants* $\alpha \in (0, 1), c > 0$ *satisfying*

$$\mathscr{L}_f \mathfrak{V}(z) \leq -c(\mathfrak{V}(z))^{\alpha}. \tag{5.10}$$

Furthermore, the settling time function satisfies

$$\mathfrak{T}(z(0)) \leq \frac{1}{c(1 - \alpha)} (\mathfrak{V}(z(0)))^{1-\alpha}. \tag{5.11}$$

If the system (5.1) has no faults and the full-order state $x^{(0 \sim n-1)}$ is measurable, then the FAS controller under the finite time-stabilization task is

$$u_{fs} = -G^{-1}(x^{(0 \sim n-1)}) \left[F(x^{(0 \sim n-1)}) + A_{0 \sim n-1} s(x^{(0 \sim n-1)}) \right], \tag{5.12}$$

where

$$\begin{cases} s(x^{(0 \sim n-1)}) = [s_1, s_2, \cdots, s_n]^{\mathsf{T}}, \\ s_i = r \operatorname{sign}(x^{(i-1)}) \| r^{n-i} x^{(i-1)} \|^{a_i}, \\ a_i = \dfrac{n a_0 - (n-1)}{(i-1) a_0 - (i-2)}, i \in [n], \end{cases} \tag{5.13}$$

$a_0 \in (\frac{n-1}{n}, 1)$, r is a positive constant greater than 1. The finite-time controller (5.12) is continuous. The proof of the finite-time stability is shown in Theorem 5.1. The finite-time FAS controller mainly consists of three characteristics: nonlinear cancellation, the parameter matrix $A_{0 \sim n-1}$ and the parameter a_i, which correspond to the global stability of the nonlinear system, the parameter tuning rule and the finite-time stability, respectively. When $a_0 \to 1, r \to 1$, the finite-time FAS stabilization controller (5.12) degenerates into the traditional asymptotic FAS controller (5.4).

Definition 5.2 (Homogeneity of Functions and Vector Fields [121, 134]) A function $V(z) : \mathbb{R}^n \to \mathbb{R}$ is homogeneous of degree γ with respect to the weights $\{\iota_i\}_{i=1}^n$, meaning that for any $\theta > 0, z = [z_1, z_2, \cdots, z_n]^{\mathsf{T}}$,

$$V(\theta^{\iota_1} z_1, \theta^{\iota_2} z_2, \cdots, \theta^{\iota_n} z_n) = \theta^{\gamma} V(z_1, z_2, \cdots, z_n). \tag{5.14}$$

A vector field $g(z) : \mathbb{R}^n \to \mathbb{R}^n$ is homogeneous of degree γ with respect to the weights $\{\iota_i\}_{i=1}^n$, meaning that for any $\theta > 0, z = [z_1, z_2, \cdots, z_n]^{\mathsf{T}}$,

$$g_i(\theta^{\iota_1} z_1, \theta^{\iota_2} z_2, \cdots, \theta^{\iota_n} z_n) = \theta^{\gamma + \iota_i} g_i(z_1, z_2, \cdots, z_n). \tag{5.15}$$

Define the following vector field $\mathscr{A}(z) : \mathbb{R}^n \to \mathbb{R}^n$,

$$\mathscr{A}(z) = [\mathscr{A}_1(z), \mathscr{A}_2(z), \cdots, \mathscr{A}_n(z)]^{\mathsf{T}}$$

$$= [z_2, z_3, \cdots, z_n, -\sum_{i=1}^{n} A_{i-1} \text{sign}(z_i) \|z_i\|^{a_i}]^{\mathsf{T}}. \tag{5.16}$$

Let $\gamma_a = a_0 - 1$, the vector field $\mathscr{A}(z)$ is homogeneous of degree γ_a with respect to the weights $\{\lambda_i\}_{i=1}^{n} \triangleq \{(i-1)a_0 - (i-2)\}_{i=1}^{n}$. A Lyapunov fucntion $V_a(z) : \mathbb{R}^n \to \mathbb{R}$ is chosen as

$$V_a(z) = z^{\mathsf{T}} P z. \tag{5.17}$$

Define a function $\mathscr{V}_a(z) : \mathbb{R}^n \to \mathbb{R}$ [85]

$$\mathscr{V}_a(z) = \int_0^{\infty} \frac{(\Psi \circ V_a)(t^{\lambda_i} z_1, t^{\lambda_2} z_2, \cdots, t^{\lambda_n} z_n)}{t^{k+1}} dt, \tag{5.18}$$

where k is a constant greater than 1, and $\circ$ represents the multiplication of mappings. $\Psi(\tau) : \mathbb{R} \to \mathbb{R}$ is a continuously differentiable function whose derivative is $\Psi'(\tau)$ satisfying

$$\Psi(\tau) = \begin{cases} 0, \tau \in (-\infty, 1], \\ 1, \tau \in [2, \infty), \end{cases} \quad \Psi'(\tau) > 0, \tau \in (1, 2). \tag{5.19}$$

The function $\Psi(\tau)$ is equal to 0 on the interval $(-\infty, 1]$ and equal to 1 on the interval $[2, \infty)$. $\Psi'(\tau)$ is greater than 0 on the interval $(1, 2)$. There are many functions meeting such requirements, e.g.,

$$\Psi(\tau) = \begin{cases} 0, \tau \leq 1, \\ 2(\tau - 1)^2, 1 < \tau \leq \frac{3}{2}, \\ 1 - 2(\tau - 2)^2, \frac{3}{2} < \tau < 2, \\ 1, \tau \geq 2. \end{cases} \tag{5.20}$$

The design objective of this function $\Psi(\tau)$ is to ensure that $\mathscr{V}_a$ is positive definite.

Lemma 5.2 *There exists a positive constant $a_* \in (0, 1)$ such that for any $a_0 \in (a_*, 1)$ and $k > 1$, $\mathscr{V}_a(z)$ is a positive definite Lyapunov function. Moreover, the following conclusions hold:*

1. *$\mathscr{L}_{\mathscr{A}} \mathscr{V}_a(z)$ is negatively definite.*
2. *$\mathscr{V}_a(z), \mathscr{L}_{\mathscr{A}} \mathscr{V}_a(z), \frac{\partial \mathscr{V}_a(z)}{\partial z_i}$ are homogeneous of degrees $k, k + \gamma_a, k - \lambda_i$, respectively, with respect to the same weights $\{\lambda_i\}_{i=1}^{n}$.*

Proof It is obvious that $\mathscr{V}_a(0) = 0, \lambda_i > 0$, and there exist two positive constants $L(z) > l(z)$ related to z such that

$$V_a(t^{\lambda_i} z_1, t^{\lambda_2} z_2, \cdots, t^{\lambda_n} z_n) \begin{cases} < 1, t \leq l(z), \\ > 2, t \geq L(z). \end{cases} \tag{5.21}$$

Combining (5.18), (5.19) and (5.21), it can be obtained that

$$\mathscr{V}_a(z) = \int_{l(z)}^{L(z)} \frac{(\Psi \circ V_a)(t^{\lambda_i} z_1, t^{\lambda_2} z_2, \cdots, t^{\lambda_n} z_n)}{t^{k+1}} dt + \frac{1}{k(L(z))^k}. \tag{5.22}$$

Therefore, $\mathscr{V}_a(z)$ is a positive definite Lyapunov function. For any $\theta > 0$, it turns out that

$$\mathscr{V}_a(\theta^{\lambda_1} z_1, \theta^{\lambda_2} z_2, \cdots, \theta^{\lambda_n} z_n)$$
$$= \int_0^\infty \frac{\theta^k (\Psi \circ V_a)((\theta t)^{\lambda_1} z_1, (\theta t)^{\lambda_2} z_2, \cdots, (\theta t)^{\lambda_n} z_n)}{(\theta t)^{k+1}} d(\theta t)$$
$$= \theta^k \mathscr{V}_a(z). \tag{5.23}$$

After taking the partial derivative of (5.23) to z_i, it follows that

$$\frac{\partial \mathscr{V}_a(\theta^{\lambda_1} z_1, \theta^{\lambda_2} z_2, \cdots, \theta^{\lambda_n} z_n)}{\partial(\theta^{\lambda_i} z_i)} = \theta^{k-\lambda_i} \frac{\partial \mathscr{V}_a(z)}{\partial z_i}. \tag{5.24}$$

Based on the homogeneity of $\mathscr{A}(z)$, it can be calculated that

$$\mathscr{A}_i(\theta^{\lambda_1} z_1, \theta^{\lambda_2} z_2, \cdots, \theta^{\lambda_n} z_n) = \theta^{\gamma_a + \lambda_i} \mathscr{A}_i(z), \tag{5.25}$$

and

$$\mathscr{L}_{\mathscr{A}} \mathscr{V}_a(\theta^{\lambda_1} z_1, \theta^{\lambda_2} z_2, \cdots, \theta^{\lambda_n} z_n) = \theta^{k+\gamma_a} \mathscr{L}_{\mathscr{A}} \mathscr{V}_a(z). \tag{5.26}$$

Define a hypersphere $\mathscr{S}_{n-1} = \{z \in \mathbb{R}^n \mid \|z\| = 1\}$ and the two continuous functions

$$\underline{V}(t) = \min_{z \in \mathscr{S}_{n-1}} V_a(t^{\lambda_1} z_1, t^{\lambda_2} z_2, \cdots, t^{\lambda_n} z_n), \tag{5.27}$$

$$\overline{V}(t) = \max_{z \in \mathscr{S}_{n-1}} V_a(t^{\lambda_1} z_1, t^{\lambda_2} z_2, \cdots, t^{\lambda_n} z_n). \tag{5.28}$$

Obviously,

$$\lim_{t \to 0} \underline{V}(t) = 0, \ \lim_{t \to \infty} \overline{V}(t) = \infty. \tag{5.29}$$

Then, there exist positive constants $L > l$ independent of z such that for any $z \in \mathscr{S}_{n-1}$,

$$V_a(t^{\lambda_i} z_1, t^{\lambda_2} z_2, \cdots, t^{\lambda_n} z_n) \begin{cases} < 1, t \le l, \\ > 2, t \ge L. \end{cases} \tag{5.30}$$

Combining (5.18), (5.19) and (5.30), it can be obtained that

$$\mathscr{V}_a(z) = \int_l^L \frac{(\Psi \circ V_a)(t^{\lambda_i}z_1, t^{\lambda_2}z_2, \cdots, t^{\lambda_n}z_n)}{t^{k+1}} dt + \frac{1}{k(L)^k}. \tag{5.31}$$

Calculating $\mathscr{L}_{\mathscr{A}}\mathscr{V}_a(z)$ yields

$$\mathscr{L}_{\mathscr{A}}\mathscr{V}_a(z) = \int_l^L \frac{(\Psi' \circ V_a)(t^{\lambda_i}z_1, t^{\lambda_2}z_2, \cdots, t^{\lambda_n}z_n)}{t^{k+1+\gamma_a}}$$
$$\cdot \left(\sum_{i=1}^n \frac{\partial V_a}{\partial z_i} \cdot \mathscr{A}_i\right) (t^{\lambda_i}z_1, t^{\lambda_2}z_2, \cdots, t^{\lambda_n}z_n) dt. \tag{5.32}$$

If $a_i = 1, i \in [n]$, it turns out that $\mathscr{A}(tz) = t\Phi_A z$,

$$\sum_{i=1}^n \left(\frac{\partial V_a}{\partial z_i} \cdot \mathscr{A}_i\right)(tz) = -t^2\|z\|^2. \tag{5.33}$$

When $a_0 \to 1$, it holds that $a_i \to 1, \lambda_i \to 1, i \in [n]$. Let

$$\Lambda(a_0) = \max_{\substack{z \in \mathscr{S}_{n-1} \\ t \in [l, L]}} \left(\sum_{i=1}^n \frac{\partial V_a}{\partial z_i} \cdot \mathscr{A}_i\right)(t^{\lambda_i}z_1, t^{\lambda_2}z_2, \cdots, t^{\lambda_n}z_n), \tag{5.34}$$

then it follows that

$$\lim_{a_0 \to 1} \Lambda(a_0) = \Lambda(1) = \max_{\substack{z \in \mathscr{S}_{n-1} \\ t \in [l, L]}} (-t^2\|z\|^2) = -l^2. \tag{5.35}$$

Thus, there exists a postive constant $a_* \in (0, 1)$ such that $\Lambda(a_0) < 0, \forall a_0 \in (a_*, 1)$. Moreover, for any $a_0 \in (a_*, 1), z \in \mathscr{S}_{n-1}$, it can be obtained that

$$\mathscr{L}_{\mathscr{A}}\mathscr{V}_a(z) \leq \Lambda(a_0) \int_l^L \frac{(\Psi' \circ V_a)(t^{\lambda_i}z_1, t^{\lambda_2}z_2, \cdots, t^{\lambda_n}z_n)}{t^{k+1+\gamma_a}} dt < 0. \tag{5.36}$$

For any $z \in \mathbb{R}^n, z \neq 0_n$, there exists a positive constant θ_* such that

$$[t^{\lambda_i}z_1, t^{\lambda_2}z_2, \cdots, t^{\lambda_n}z_n]^\mathsf{T} \in \mathscr{S}_{n-1}. \tag{5.37}$$

Based on (5.26) and (5.36), for any $z \in \mathbb{R}^n, z \neq 0_n$, it follows that

$$\mathscr{L}_{\mathscr{A}}\mathscr{V}_a(z) = \frac{1}{\theta_*^{k+\gamma_a}} \mathscr{L}_{\mathscr{A}}\mathscr{V}_a(\theta^{\lambda_1}z_1, \theta^{\lambda_2}z_2, \cdots, \theta^{\lambda_n}z_n) < 0, \tag{5.38}$$

and $\mathscr{L}_{\mathscr{A}}\mathscr{V}(\mathbf{0}_n) = 0$. Lemma 5.2 has been proved.

Lemma 5.3 ([10]) *Suppose two continuous functions $V_1(z), V_2(z) : \mathbb{R}^n \to \mathbb{R}$ are homogeneous of degrees v_1, v_2, respectively, with respect to the same weights, and $V_1(z)$ is a positive definite function, then for any $z \in \mathbb{R}^n$, the following conclusion holds:*

$$\left(\min_{z \in \{z \in \mathbb{R}^n | V_1(z)=1\}} V_2(z) \right) \cdot (V_1(z))^{\frac{v_2}{v_1}} \le V_2(z)$$
$$\le \left(\max_{z \in \{z \in \mathbb{R}^n | V_1(z)=1\}} V_2(z) \right) \cdot (V_1(z))^{\frac{v_2}{v_1}}. \tag{5.39}$$

Theorem 5.1 *Consider the fault-free system (5.12) under the finite-time controller (5.1), for any $a_0 \in (a_*, 1)$ and $k > 1$, the FAS (5.1) is globally finite-time stable, i.e.,*

$$x^{(0 \sim n-1)}(t) = 0, \forall t \ge T, \tag{5.40}$$

where T is a positive constant related to $x^{(0 \sim n-1)}(0)$.

Proof Define

$$\zeta = [\zeta_1, \zeta_2, \cdots, \zeta_n]^\mathsf{T} = [r^{n-1}x, r^{n-2}\dot{x}, \cdots, x^{(n-1)}]^\mathsf{T}, \tag{5.41}$$

and the dynamics of ζ is

$$\begin{cases} \dot{\zeta}_i = r\zeta_{i+1}, i \in [n-1], \\ \dot{\zeta}_n = -r \sum_{i=1}^{n} A_{i-1} \operatorname{sign}(\zeta_i) \|\zeta_i\|^{a_i}. \end{cases} \tag{5.42}$$

The derivative of $\mathscr{V}_a(\zeta)$ is

$$\frac{d\mathscr{V}_a(\zeta)}{dt} = \sum_{i=1}^{n} \dot{\zeta}_i \frac{\partial \mathscr{V}_a(\zeta)}{\partial \zeta_i} = r\mathscr{L}_{\mathscr{A}}\mathscr{V}_a(\zeta). \tag{5.43}$$

Based on Lemmas 5.2 and 5.3, there exists a positive constant ρ such that

$$\mathscr{L}_{\mathscr{A}}\mathscr{V}_a(\zeta) \le -\rho \, (\mathscr{V}_a(\zeta))^{\frac{k+\gamma_a}{k}}. \tag{5.44}$$

It follows that

$$\frac{d\mathscr{V}_a(\zeta)}{dt} \le -r\rho \, (\mathscr{V}_a(\zeta))^{\frac{k+\gamma_a}{k}}. \tag{5.45}$$

Based on Theorem 5.1, there exists a positive constant

$$T_0 = -\frac{k}{r\rho\gamma_a}(\mathcal{V}_a(\zeta(0)))^{-\frac{\gamma_a}{k}} \tag{5.46}$$

such that

$$\zeta(t) = 0, \forall t \geq T_0. \tag{5.47}$$

Theorem 5.1 has been proved.

Remark 5.2 The characteristics of the finite-time FAS controller lies in the partial retention and partial innovation of the parametric structure of FASs. It is necessary to retain the parameters that are easy to tune while improving the control performance. On the basis of fully measurable full-order states, the structure of the finite-time FAS stabilization controller (5.12) provides a reference for the observation-based FTFTC design in the next section.

5.3 Design of Fault-Tolerant Control

If the FAS model (5.1) is fault-free and the system state $x^{(0\sim n-1)}$ is completely measurable, the traditional asymptotic tracking controller is

$$u_{at} = -G^{-1}(x^{(0\sim n-1)})\left[F(x^{(0\sim n-1)}) - x_r^{(n)} + A_{0\sim n-1}(x^{(0\sim n-1)} - x_r^{(0\sim n-1)})\right], \tag{5.48}$$

where $x_r \in \mathbb{R}$ is the reference signal satisfying $\sup_{t\geq 0}\|x_r^{(0\sim n)}\| < \bar{x}_r, \bar{x}_r > 0$. Then the dynamics of tracking error $e = x - x_r$ is

$$\dot{e}^{(0\sim n-1)} = \Phi_A e^{(0\sim n-1)}. \tag{5.49}$$

In this case, the Hurwitz matrix Φ_A can guarantee the performance of global asymptotic tracking. Under the finite-time tracking control task, the corresponding parameterized FAS controller is

$$u = -G^{-1}(x^{(0\sim n-1)})\left[F(x^{(0\sim n-1)}) - x_r^{(n)} + A_{0\sim n-1}s_e(x^{(0\sim n-1)} - x_r^{(0\sim n-1)})\right], \tag{5.50}$$

where

$$\begin{cases} s_e(e^{(0\sim n-1)}) = [s_{e1}, s_{e2}, \cdots, s_{en}]^{\mathsf{T}}, \\ s_{ei} = r\,\mathrm{sign}(e^{(i-1)})\|r^{n-i}e^{(i-1)}\|^{a_i}. \end{cases} \tag{5.51}$$

The corresponding parameters are shown as (5.13), especially when $a_0 \rightarrow 1, r \rightarrow 1$, the finite-time tracking controller (5.50) degenerates into the traditional asymptotic tracking controller (5.48). Unfortunately, when the system state $x^{(0 \sim n-1)}$ cannot be directly measured and the FAS model (5.1) is faulty, the finite-time controller (5.50) is not easy to set. Therefore, a finite-time observer is designed as

$$
\begin{cases}
\dot{\hat{x}}_i = \hat{x}_{i+1} + \dfrac{B_{i-1}}{\epsilon^{n-i}} \operatorname{sign}(\varepsilon_1) \|\varepsilon_1\|^{b_i}, i \in [n-1], \\
\dot{\hat{x}}_n = F(\hat{X}) + G(\hat{X})u + \hat{\delta} + B_{n-1} \operatorname{sign}(\varepsilon_1) \|\varepsilon_1\|^{b_n}, \\
\dot{\hat{\delta}} = B_n \epsilon \operatorname{sign}(\varepsilon_1) \|\varepsilon_1\|^{b_{n+1}},
\end{cases}
\tag{5.52}
$$

where $\hat{X} = [\hat{x}_1, \hat{x}_2, \cdots, \hat{x}_n]^\mathsf{T}$ is the observation of $X = [x_1, x_2, \cdots, x_n]^\mathsf{T} \triangleq x^{(0 \sim n-1)}$, $\hat{\delta}$ is the fault estimation, ϵ is a constant greater than 1, $\varepsilon_i = \epsilon^{n+1-i}(x_i - \hat{x}_i)$, $b_i = ib_0 - (i-1), i \in [n]$, and b_0 is a constant on the interval $(\frac{n}{n+1}, 1)$. The parameters $B_{i-1}, i \in [n+1]$ can shape a matrix.

$$
\Phi_B = \begin{bmatrix}
-B_0 & 1 & 0 & \cdots & 0 \\
-B_1 & 0 & 1 & \cdots & 0 \\
\vdots & & & \ddots & \\
-B_{n-1} & 0 & 0 & \cdots & 1 \\
-B_n & 0 & 0 & \cdots & 0
\end{bmatrix}
\tag{5.53}
$$

There exists a positive definite matrix $P_B \in \mathbb{R}^{(n+1) \times (n+1)}$ that is a solution to the following Lyapunov equation

$$
P_B \Phi_B + \Phi_B^\mathsf{T} P_B = -I_{n+1}.
\tag{5.54}
$$

Furthermore, in order to meet the amplitude requirements of the actual actuator, the saturation strategy ts adopted to limit the fault-tolerant controller:

$$
u = -\operatorname{Sat}_M(G^{-1}(\hat{X})[F(\hat{X}) + \hat{\delta} - x_r^{(n)} + A_{0 \sim n-1}\hat{s}_e(\hat{X} - x_r^{(0 \sim n-1)})]),
\tag{5.55}
$$

where

$$
\begin{cases}
\hat{s}_e(\hat{X} - x_r^{(0 \sim n-1)}) = [\hat{s}_{e1}, \hat{s}_{e2}, \cdots, \hat{s}_{en}]^\mathsf{T}, \\
\hat{s}_{ei} = r \operatorname{sign}(\hat{x}_i - x_r^{(i-1)}) \|r^{n-i}(\hat{x}_i - x_r^{(i-1)})\|^{a_i},
\end{cases}
\tag{5.56}
$$

The sign function can be chosen as $\operatorname{Sat}_M(\cdot) \triangleq \min\{\max\{\cdot, -M\}, \mathscr{M}\}$. Under normal circumstances, M can be conservatively determined through mechanism knowledge and computer simulation. The combination of observer (5.52) and controller (5.55) can ensure the finite-time stability of the tracking error system.

Define the vector fields $\mathscr{B}(\mathfrak{z}) : \mathbb{R}^{n+1} \rightarrow \mathbb{R}^{n+1}$,

$$\mathscr{B}(\mathfrak{z}) = [\mathscr{B}_1(\mathfrak{z}), \mathscr{B}_2(\mathfrak{z}), \cdots, \mathscr{B}_{n+1}(\mathfrak{z})]^{\mathsf{T}} = \begin{bmatrix} \mathfrak{z}_2 - B_0 \mathrm{sign}(\mathfrak{z}_1) \|\mathfrak{z}_1\|^{b_1} \\ \mathfrak{z}_3 - B_1 \mathrm{sign}(\mathfrak{z}_1) \|\mathfrak{z}_1\|^{b_2} \\ \vdots \\ \mathfrak{z}_{n+1} - B_{n-1} \mathrm{sign}(\mathfrak{z}_1) \|\mathfrak{z}_1\|^{b_n} \\ -B_n \mathrm{sign}(\mathfrak{z}_1) \|\mathfrak{z}_1\|^{b_{n+1}} \end{bmatrix},$$

(5.57)

where $\mathfrak{z} = [\mathfrak{z}_1, \mathfrak{z}_2, \cdots, \mathfrak{z}_{n+1}]^{\mathsf{T}}$. Let $\gamma_b = b_0 - 1$, $\mathscr{B}(\mathfrak{z})$ is homogeneous of degrees γ_b with respect to the weights $\{\mu_i\}_{i=1}^{n+1} \triangleq \{(i-1)b_0 - (i-2)\}_{i=1}^{n+1}$. A Lyapunov function $V(\mathfrak{z}) : \mathbb{R}^{n+1} \to \mathbb{R}$ is chosen as

$$V_b(z) = \mathfrak{z}^{\mathsf{T}} P_B \mathfrak{z}. \tag{5.58}$$

Define $\mathscr{V}_b(\mathfrak{z}) : \mathbb{R}^{n+1} \to \mathbb{R}$,

$$\mathscr{V}_b(\mathfrak{z}) = \int_0^{\infty} \frac{(\Psi \circ V_b)(t^{\mu_i} \mathfrak{z}_1, t^{\mu_2} \mathfrak{z}_2, \cdots, t^{\mu_{n+1}} \mathfrak{z}_{n+1})}{t^{k+1}} dt. \tag{5.59}$$

Lemma 5.4 *There exists a positive scalar $b_* \in (0, 1)$ such that for any $b_0 \in (b_*, 1)$ and $k > 1$, $\mathscr{V}_b(\mathfrak{z})$ is a positive definite Lyapunov function, and the following conditions hold:*

1. *$\mathscr{L}_{\mathscr{B}} \mathscr{V}_b(\mathfrak{z})$ is negatively definite.*
2. *$\mathscr{V}_b(\mathfrak{z}), \mathscr{L}_{\mathscr{B}} \mathscr{V}_b(\mathfrak{z}), \frac{\partial \mathscr{V}_b(\mathfrak{z})}{\partial \mathfrak{z}_i}$ are homogeneous of degrees $k, k + \gamma_b, k - \mu_i$, respectively, with respect to the same weights $\{\mu_i\}_{i=1}^{n+1}$.*

Proof The proof process is similar to Lemma 5.2.

Theorem 5.2 *For FAS model (5.1) under the finite-time observer (5.52) and the finite-time controller (5.55), there exist $\bar{r}_*, \bar{\epsilon}_* > 0$ such that for any $a_0 \in (a_*, 1), b_0 \in (b_*, 1), r > \bar{r}_*, \epsilon > \bar{\epsilon}_*, k > 1$, the observation error and tracking error are finite-time stable, i.e.,*

$$\begin{cases} \hat{X}(t) = x^{(0 \sim n-1)}(t), \forall t \geq T_1, \\ \hat{\delta}(t) = \delta(t), \forall t \geq T_1, \\ x^{(0 \sim n-1)}(t) = x_r^{(0 \sim n-1)}(t), \forall t \geq T_2, \end{cases} \tag{5.60}$$

where T_1, T_2 are two constants related to the initial state.

Proof Let

$$\varepsilon = [\varepsilon_1, \varepsilon_2, \cdots, \varepsilon_n, \varepsilon_{n+1}]^{\mathsf{T}} = [\epsilon^n(x - \hat{x}_1), \epsilon^{n-1}(\dot{x} - \hat{x}_2), \cdots, \delta - \hat{\delta}]^{\mathsf{T}},$$

(5.61)

the derivative of ε is

$$
\begin{cases}
\dot{\varepsilon}_i = \epsilon(\varepsilon_{i+1} - B_{i-1}\mathrm{sign}(\varepsilon_1)\|\varepsilon_1\|^{b_i}), i \in [n-1], \\
\dot{\varepsilon}_n = \epsilon(\varepsilon_{n+1} - B_{n-1}\mathrm{sign}(\varepsilon_1)\|\varepsilon_1\|^{b_n}) + \epsilon\varpi, \\
\dot{\varepsilon}_{n+1} = -\epsilon B_n\mathrm{sign}(\varepsilon_1)\|\varepsilon_1\|^{b_{n+1}} + \dot{\delta},
\end{cases}
\tag{5.62}
$$

where

$$
\varpi = F(x^{(0\sim n-1)}) - F(\hat{X}) + (G(x^{(0\sim n-1)}) - G(\hat{X}))u. \tag{5.63}
$$

Based on Assumption 5.1, it turns out that

$$
\|F(x^{(0\sim n-1)}) - F(\hat{X})\| \leq \bar{F}\|x^{(0\sim n-1)} - \hat{X}\| \leq \frac{1}{\epsilon}\bar{F}\sum_{i=1}^{n}\|\varepsilon_i\|, \tag{5.64}
$$

$$
\|F(x^{(0\sim n-1)}) - F(\hat{X})\| \leq \bar{G}\|x^{(0\sim n-1)} - \hat{X}\| \leq \frac{1}{\epsilon}\bar{G}\sum_{i=1}^{n}\|\varepsilon_i\|. \tag{5.65}
$$

Then, it follows that

$$
\|\varpi\| \leq \frac{1}{\epsilon}(\bar{F} + M\bar{G})\sum_{i=1}^{n}\|\varepsilon_i\| = \frac{N}{\epsilon}\sum_{i=1}^{n}\|\varepsilon_i\|. \tag{5.66}
$$

where $N = \bar{F} + M\bar{G}$. Obviouly, $V_i(\varepsilon) = \|\varepsilon_i\|, i \in [n+1]$ are homogeneous of degree μ_i with respect to the weights $\{\mu_i\}_{i=1}^{n+1}$. Based on Lemma 5.3, $\mathcal{V}_b(\varepsilon)$ is homogeneous of degree k with respect to $\{\mu_i\}_{i=1}^{n+1}$. Based on Lemma 5.4, there exists $\rho_{1,i} = \max_{\varepsilon \in \{\varepsilon \in \mathbb{R}^{n+1} | \mathcal{V}_b(\varepsilon)=1\}} V_i(\varepsilon), i \in [n+1]$ such that

$$
\|\varepsilon_i\| \leq \rho_{1,i}(\mathcal{V}_b(\varepsilon))^{\frac{\mu_i}{k}}. \tag{5.67}
$$

Then, there exist $\rho_1 = \max_{i \in [n+1]}\{\rho_{1,i}\}$ such that

$$
\|\varpi\| \leq \frac{\rho_1 N}{\epsilon}\sum_{i=1}^{n}(\mathcal{V}_b(\varepsilon))^{\frac{\mu_i}{k}}. \tag{5.68}
$$

Moreover, the derivative of $\mathcal{V}_b(\varepsilon)$ is

$$
\frac{d\mathcal{V}_b(\varepsilon)}{dt} = \epsilon\mathcal{L}_{\mathscr{B}}\mathcal{V}_b(\varepsilon) + \epsilon\varpi\frac{\partial\mathcal{V}_b(\varepsilon)}{\partial\varepsilon_n} + \dot{\delta}\frac{\partial\mathcal{V}_b(\varepsilon)}{\partial\varepsilon_{n+1}}
$$

$$\leq -\epsilon \rho_2 (\mathcal{V}_b(\varepsilon))^{\frac{k+\gamma_b}{k}} + \rho_1 \rho_3 N \sum_{i=1}^{n} (\mathcal{V}_b(\varepsilon))^{\frac{k-\mu_n+\mu_i}{k}} + \bar{\delta}\rho_3 (\mathcal{V}_b(\varepsilon))^{\frac{k-\mu_{n+1}}{k}},$$

$$\tag{5.69}$$

where ρ_2, ρ_3 come from (5.70). Lemma 5.4 means that $\mathcal{V}_b(\varepsilon)$, $\mathcal{L}_{\mathcal{B}}\mathcal{V}_b(\varepsilon)$, $\frac{\partial \mathcal{V}_b(\varepsilon)}{\partial \varepsilon_i}$ are homogeneous of degrees k, $k + \gamma_b$, $k - \mu_i$, respectively, with respective to the same weights. Since $\mathcal{V}_b(\varepsilon)$, $\mathcal{L}_{\mathcal{B}}\mathcal{V}_b(\varepsilon)$, $\frac{\partial \mathcal{V}_b(\varepsilon)}{\partial \varepsilon_i}$ meets Lemma 5.3, there exist ρ_2, ρ_3 such that

$$\mathcal{L}_{\mathcal{B}}\mathcal{V}_b(\varepsilon) \leq -\rho_2 (\mathcal{V}_b(\varepsilon))^{\frac{k+\gamma_b}{k}}, \quad \frac{\partial \mathcal{V}_b(\varepsilon)}{\partial \varepsilon_i} \leq \rho_3 (\mathcal{V}_b(\varepsilon))^{\frac{k-\mu_i}{k}}. \tag{5.70}$$

Define a positive definite function $\mathcal{V}_b(\mathfrak{z}) : \mathbb{R}^{n+1} \to \mathbb{R}$,

$$\mathcal{V}_b(\mathfrak{z}) = \|\mathfrak{z}_1\|^{\frac{k}{\mu_1}} + \|\mathfrak{z}_2\|^{\frac{k}{\mu_2}} + \cdots + \|\mathfrak{z}_{n+1}\|^{\frac{k}{\mu_{n+1}}}, \tag{5.71}$$

and $\mathcal{V}_b(\mathfrak{z})$ is homogeneous of degree k with respect to $\{\mu_i\}_{i=1}^{n+1}$. Based on Lemmas 5.3 and 5.4, it can be obtained that

$$\mathcal{V}_b(\mathfrak{z}) \leq \sigma_b \mathcal{V}_b(\mathfrak{z}), \tag{5.72}$$

where $\sigma_b > 0$. The initial value $\varepsilon(0)$ satisfies

$$\mathcal{V}_b(\varepsilon(0)) \leq \sigma_b \mathcal{V}_b(\epsilon^n (x(0) - \hat{x}_1(0)), \cdots, \delta(0) - \hat{\delta}(0))$$

$$= \sigma_b \sum_{i=1}^{n} \|\epsilon^{n+1-i}(x^{(i-1)}(0) - \hat{x}_i(0))\|^{\frac{k}{\mu_i}} + \sigma_b \|\varepsilon(0)\|^{\frac{k}{\mu_{n+1}}}$$

$$\leq \tau \sum_{i=1}^{n+1} \epsilon^{\frac{(n+1-i)k}{\mu_i}} \leq \tau \epsilon^{nk}, \tag{5.73}$$

where

$$\tau = \max \left\{ 1, \sigma_b \sum_{i=1}^{n} \|x^{(i-1)}(0) - \hat{x}_i(0)\|^{\frac{k}{\mu_i}}, \sigma_b \|\varepsilon(0)\|^{\frac{k}{\mu_{n+1}}} \right\}. \tag{5.74}$$

Define two compact sets

$$\Omega_1 = \{\mathfrak{z} \in \mathbb{R}^{n+1} \mid \mathcal{V}_b(\mathfrak{z}) \leq 1\}, \tag{5.75}$$

$$\Omega_2 = \{\mathfrak{z} \in \mathbb{R}^{n+1} \mid \mathcal{V}_b(\mathfrak{z}) \leq \tau \epsilon^{nk}\}, \tag{5.76}$$

and it turns out that $\varepsilon(0) \in \Omega_2$, $\Omega_1 \subset \Omega_2$. Let

$$\bar{\epsilon}_1 = \left[\frac{\rho_1 \rho_3 N n(n+1) + 2\bar{\delta}\rho_3}{\rho_2} \right]^{\frac{1}{1+n^2\gamma_b}} \tau^{\frac{-n\gamma_b}{k(1+n^2\gamma_b)}}, \tag{5.77}$$

for any $\epsilon > \bar{\epsilon}_1$, it follows that

$$\epsilon\rho_2 \geq [\rho_1\rho_3 Nn(n+1) + 2\bar{\delta}\rho_3]\tau^{\frac{-n\gamma_b}{k}} \epsilon^{-n^2\gamma_b}, \tag{5.78}$$

If $\varepsilon(t) \in \Omega_2 - \Omega_1$, i.e., $1 < \mathcal{V}_b(\varepsilon) \leq \tau\epsilon^{nk}$, for any $\epsilon > \bar{\epsilon}_1$, it can be obtained that

$$\epsilon\rho_2 \geq [\rho_1\rho_3 Nn(n+1) + 2\bar{\delta}\rho_3](\mathcal{V}_b(\varepsilon))^{\frac{-n\gamma_b}{k}}$$
$$= [\rho_1\rho_3 Nn(n+1) + 2\bar{\delta}\rho_3](\mathcal{V}_b(\varepsilon))^{\frac{k-\mu_n+\mu_1-(k+\gamma_b)}{k}}. \tag{5.79}$$

Based on (5.79), for any $\varepsilon(t) \in \Omega_2 - \Omega_1$, it turns out that

$$\frac{d\mathcal{V}_b(\varepsilon)}{dt} \leq -\epsilon\rho_2(\mathcal{V}_b(\varepsilon))^{\frac{k+\gamma_b}{k}} + \bar{\delta}\rho_3(\mathcal{V}_b(\varepsilon))^{\frac{k-\mu_n+\mu_1}{k}}$$
$$+ \frac{\rho_1\rho_3 Nn(n+1)}{2}(\mathcal{V}_b(\varepsilon))^{\frac{k-\mu_n+\mu_1}{k}}$$
$$\leq -\frac{\epsilon}{2}\rho_2(\mathcal{V}_b(\varepsilon))^{\frac{k+\gamma_b}{k}}. \tag{5.80}$$

This means $\forall \epsilon > \bar{\epsilon}_1$, $\varepsilon(t) \in \Omega_2$.

Consider a differentiable function

$$\dot{\Upsilon}(t) = -\frac{\epsilon\rho_2}{2}(\Upsilon(t))^{\frac{k+\gamma_b}{k}}, \quad \Upsilon(0) = \mathcal{V}_b(\varepsilon(0)). \tag{5.81}$$

the solution to (5.81) is

$$\Upsilon(t) = \begin{cases} \left[\frac{\epsilon\rho_2\gamma_b}{2k}t + (\mathcal{V}_b(\varepsilon(0)))^{\frac{-\gamma_b}{k}} \right]^{\frac{k}{-\gamma_b}}, & 0 \leq t < t_1, \\ 0, & t \geq t_1, \end{cases} \tag{5.82}$$

where

$$t_1 = \frac{2k}{-\epsilon\rho_2\gamma_b}(\mathcal{V}_b(\varepsilon(0)))^{\frac{-\gamma_b}{k}} \leq \frac{2k}{-\epsilon\rho_2\gamma_b}\tau^{\frac{-\gamma_b}{k}}\left(\frac{1}{\epsilon}\right)^{n\gamma_b}$$
$$= \frac{2k}{-\rho_2\gamma_b}\tau^{\frac{-\gamma_b}{k}}\left(\frac{1}{\epsilon}\right)^{1+n(b_0-1)}. \tag{5.83}$$

Obviously, when $\epsilon \to \infty$, it holds that $t_1 \to 0$. Comparing (5.80) with (5.81), there exist $\bar{\epsilon}_2 > \bar{\epsilon}_1, t_2 > t_1$ such that

$$\mathcal{V}_b(\varepsilon) \le 1, \forall t \ge t_2, \epsilon > \bar{\epsilon}_2. \tag{5.84}$$

When $b_0 \in (\frac{n}{n+1}, 1)$, it follows that

$$k - \mu_n + \mu_i \ge k - \mu_n + \mu_{n+1} = k + \gamma_b, i \in [n], k - \mu_{n+1} \ge k + \gamma_b. \tag{5.85}$$

Moreover, for any $t \ge t_2, \epsilon > \bar{\epsilon}_2$, the derivative of $\mathcal{V}_b(\varepsilon)$ satisfies

$$\frac{d\mathcal{V}_b(\varepsilon)}{dt} \le -\epsilon\rho_2(\mathcal{V}_b(\varepsilon))^{\frac{k+\gamma_b}{k}} + \bar{\delta}\rho_3(\mathcal{V}_b(\varepsilon))^{\frac{k+\gamma_b}{k}} + \frac{\rho_1\rho_3 Nn(n+1)}{2}(\mathcal{V}_b(\varepsilon))^{\frac{k+\gamma_b}{k}}$$

$$\le -\frac{\epsilon}{2}\rho_2(\mathcal{V}_b(\varepsilon))^{\frac{k+\gamma_b}{k}}. \tag{5.86}$$

Therefore, there exist $\epsilon_* > \bar{\epsilon}_2, T_1 > t_2$ such that

$$\mathcal{V}_b(\varepsilon(t)) = 0, \forall \epsilon > \epsilon_*, t \ge T_1. \tag{5.87}$$

Then, it follows that

$$\hat{X}(t) = x^{(0\sim n-1)}(t), \hat{\delta}(t) = \delta(t), \forall \epsilon > \epsilon_*, t \ge T_1. \tag{5.88}$$

Let

$$w = [w_1, w_2, \cdots, w_n]^{\mathsf{T}} = [r^{n-1}e, r^{n-2}\dot{e}, \cdots, e^{(n-1)}]^{\mathsf{T}}, \tag{5.89}$$

the derivative of w is

$$\begin{cases} \dot{w}_i = rw_{i+1}, i \in [n-1], \\ \dot{w}_n = F(x^{(0\sim n-1)}) + G(x^{(0\sim n-1)})u + \delta - x_r^{(n)}. \end{cases} \tag{5.90}$$

Define a positive definite function $\mathcal{V}_a(z) : \mathbb{R}^n \to \mathbb{R}$,

$$\mathcal{V}_a(z) = \|z_1\|^{\frac{k}{\lambda_1}} + \|z_2\|^{\frac{k}{\lambda_2}} + \cdots + \|z_n\|^{\frac{k}{\lambda_n}}, \tag{5.91}$$

$\mathcal{V}_a(z)$ is homogeneous of degree k with respect to the weights $\{\lambda_i\}_{i=1}^n$. Based on Lemma 5.2 and Lemma 5.3, it can be obtained that

$$\mathcal{V}_a(z) \le \sigma_a \mathcal{V}_a(z), \tag{5.92}$$

where σ_a is a positive constant. Therfore, the initial value $w(0)$ satisfies

$$\mathscr{V}_a(w(0)) \leq \sigma_a \mathscr{V}(r^{n-1}e(0), r^{n-2}\dot{e}(0), \cdots, e^{(n-1)}(0))$$

$$= \sigma_1 \sum_{i=1}^{n} \left\| r^{n-i}e^{(i-1)}(0) \right\|^{\frac{k}{\lambda_i}}$$

$$\leq \sigma_1 (1+\kappa)^{\frac{k}{\lambda_n}} \sum_{i=1}^{n} r^{\frac{(n-i)k}{(i-1)a_0-(i-2)}}$$

$$\leq n\sigma_1 (1+\kappa)^{\frac{k}{\lambda_n}} r^{nk}, \tag{5.93}$$

where $\kappa = \max_{i=1,2,\cdots,n} |e^{(i-1)}(0)|$.

Consider two compact sets

$$\Omega_3 = \{z \in \mathbb{R}^n \mid \mathscr{V}_a(z) \leq n\sigma_1 (1+\kappa)^{\frac{k}{\lambda_n}} r^{nk}\}, \tag{5.94}$$

$$\Omega_4 = \{z \in \mathbb{R}^n \mid \mathscr{V}_a(z) \leq n\sigma_1 (1+\kappa)^{\frac{k}{\lambda_n}} r^{nk} + 1\}, \tag{5.95}$$

it turns out that $w(0) \in \Omega_3$. Let

$$\varsigma = \max_{z \in \Omega_4} \left\{ \|\tilde{z}_i\| + \|F(\tilde{z})\| + \|G(\tilde{z})\|M + \bar{\delta} + \bar{x}_r \right\}, \quad t_3 = 1/\varsigma, \tag{5.96}$$

where

$$\tilde{z} = [\frac{z_1}{r^{n-1}} + x_r, \frac{z_2}{r^{n-2}} + \dot{x}_r, \cdots, z_n + x_r^{(n-1)}]^{\mathsf{T}}. \tag{5.97}$$

It follows that

$$w(t) \in \Omega_3, 0 \leq t \leq t_3. \tag{5.98}$$

Otherwise, there exists $t_3^* < t_3$ such that

$$w(t) \in \Omega_3, 0 \leq t < t_3^*, \ w(t_0^*) \in \Omega_4 - \Omega_3. \tag{5.99}$$

Based on (5.90), for $i \in [n]$, it turns out that

$$\|e^{(i-1)}(t_3^*)\| \leq \|e^{(i-1)}(0)\| + \varsigma t_3^* < \kappa + 1. \tag{5.100}$$

Therefore,

$$\mathscr{V}_a(w(t_3^*)) \leq n\sigma_a (1+\kappa)^{\frac{k}{\lambda_n}} r^{nk}. \tag{5.101}$$

This contradicts (5.99). Furthermore, it can be proved by contradiction that there exists a positive constant $\bar{r}_*$ such that

$$w(t) \in \Omega_3, \forall r > \bar{r}_*, t \geq 0. \tag{5.102}$$

Otherwise, there exists $r > \bar{r}_*, t_5 > t_4 > t_3$ such that

$$w(t_4) \in \partial\Omega_3, w(t) \in \Omega_4 - \Omega_3, t_4 < t \leq t_5, \tag{5.103}$$

where $\partial\Omega_3$ represents the boundary of Ω_3. Based on (5.88), there exists $\bar{\epsilon}_* > 0$ such that $\hat{X}(t) = x^{(0 \sim n-1)}(t), \hat{\delta}(t) = \delta(t), \forall \epsilon > \bar{\epsilon}_*, t_3 \leq t \leq t_5$. To avoid saturation in the interval $[t_3, t_5]$, the parameter M should satisfy

$$M \geq \sup_{e^{(0 \sim n-1)} \in \Omega_4} \left\{ \left\| G^{-1}(x^{(0 \sim n-1)})[F(x^{(0 \sim n-1)}) - x_r^{(n)} + A_{0 \sim n-1} S_e(e^{(0 \sim n-1)})] \right\| \right.$$

$$\left. + \bar{\delta} \right\}, \tag{5.104}$$

where $x^{(0 \sim n-1)} = e^{(0 \sim n-1)} + x_r^{(0 \sim n-1)}$. Thus, for $\epsilon > \bar{\epsilon}_*, r > \bar{r}_*, t \in [t_3, t_5]$, the dynamic equations are

$$\begin{cases} \dot{w}_i = r w_{i+1}, i \in [n-1], \\ \dot{w}_n = -r \sum_{i=1}^n A_{i-1} \text{sign}(w_i) \|w_i\|^{a_i}. \end{cases} \tag{5.105}$$

Similar to the proof of Theorem 5.1, the following conclusion holds:

$$\frac{d\mathcal{V}_a(w)}{dt} \leq -r\rho \left(\mathcal{V}_a(w)\right)^{\frac{k+\gamma_a}{k}}, r > \bar{r}_*, t \in [t_3, t_5]. \tag{5.106}$$

This contradicts (5.103). Therefore, there exists $r > \bar{r}_*$ such that

$$w(t) \in \Omega_3, \forall \epsilon > \bar{\epsilon}_*, r > \bar{r}_*, t \geq 0. \tag{5.107}$$

If $w(t) \in \Omega_3$, (5.107) holds. Based on Lemma 5.1 and $w(0) \in \Omega_3$, the tracking error is finite-time stable, i.e., there exists $T_2 > T_1$ such that $w(t) = 0, x^{(0 \sim n-1)}(t) = x_r^{(0 \sim n-1)}(t), \forall t \geq T_2$. Theorem 5.2 has been proved.

Remark 5.3 Most existing FAS approaches, i.e, robust adaptive control [26], predictive control [63], event-triggered control [76], etc. can only yield asymptotic stability results. The proposed FTFTC approach is a parametric finite-time control scheme different from the traditional structure. It not only retains the advantages of the FAS theory but also improves the system response speed. In the field of fault-tolerant control (FTC), the FTFTC of most nonlinear systems are first-order state-space methods. These methods are often state-coupled, and their controller structures are complex and the parameter tuning is difficult. For instance, the parameters of the FTFTC based on neural networks are difficult to adjust [117], and the FTFTC based on the backstepping method will face the problem of differential explosion [60]. The FTFTC approach proposed in this chapter is based on the FAS theory. Decoupled controller design and standard parameter design are its obvious advantages. Furthermore, compared with the FTC design that adopted the FAS approach in the previous chapters, the fault compensation speed in this chapter is faster.

5.4 Simulation Results of Real-World Models

5.4.1 A Stabilization Experiment of a Measurable System

Consider the measurable FAS model

$$\dddot{x} = x^2 + \dot{x}\ddot{x} + (1 + x^2 + \dot{x}^3)u, \tag{5.108}$$

where $F(x^{(0\sim2)}) = x^2 + \dot{x}\ddot{x}$, $G(x^{(0\sim2)}) = 1 + x^2 + \dot{x}^3$. The finite-time FAS controller is designed as

$$u_{fs} = -G^{-1}(x^{(0\sim2)})\left[F(x^{(0\sim2)}) + A_{0\sim2}s(x^{(0\sim2)})\right]. \tag{5.109}$$

Correspondingly, the asymptotic FAS controller is designed as

$$u_{as} = -G^{-1}(x^{(0\sim2)})(F(x^{(0\sim2)}) + A_{0\sim2}x^{(0\sim2)}). \tag{5.110}$$

To facilitate the comparison of the stabilizing controllers, the parameters $A_{0\sim2} = [1, 1.5, 1.5]^\mathsf{T}$ are kept consistent, the initial state is set as $x^{(0\sim2)}(0) = [1, 0, -1]^\mathsf{T}$, and other parameters are set as $r = 1.5$, $a_0 = 0.85$. The results of the comparative experiments are shown in Fig. 5.1. Under the controller (5.109), the FAS (5.108) can be stabilized within a finite time. The stabilization speed under the finite-time controller (5.109) is significantly faster than that under the asymptotic controller (5.110).

5.4.2 An FTC Experiment of an Unmeasurable System

To illustrate the effectiveness of the proposed FTFTC framework, consider a single-link robot model

$$\begin{cases} \ddot{q} + 10\sin q = \tau_q + \delta, \\ y = q, \end{cases} \tag{5.111}$$

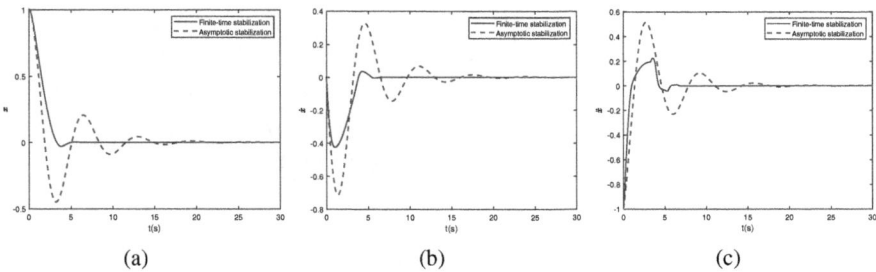

Fig. 5.1 The signal trajectories in the stabilization experiment

where q, τ_q, δ represent the angle, input torque, and fault signal, respectively. The actuator fault signal injected in the fault-tolerant tracking control experiment is

$$\delta(t) = \begin{cases} 0, 0 \leq t < 15\text{s}, \\ 2 + \sin(t - 15), t \geq 15\text{s}. \end{cases} \tag{5.112}$$

The reference signal is $q_r = \sin t$. In the finite-time fault-tolerant tracking control experiment, the finite-time observer is designed as follows:

$$\begin{cases} \dfrac{d\hat{q}}{dt} = \hat{\dot{q}} + \dfrac{2}{\epsilon}\text{sign}(y - \hat{q})\|\epsilon^2(y - \hat{q})\|^{b_1}, \\ \dfrac{d\hat{\dot{q}}}{dt} = -10\sin\hat{q} + \hat{\delta} + \tau_q + 4\text{sign}(y - \hat{q})\|\epsilon^2(y - \hat{q})\|^{b_2}, \\ \dfrac{d\hat{\delta}}{dt} = 3\epsilon\text{sign}(y - \hat{q})\|\epsilon^2(y - \hat{q})\|^{b_3}, \end{cases} \tag{5.113}$$

where $\hat{q}, \hat{\dot{q}}, \hat{\delta}$ are the estimations. The corresponding FTFTC controller is

$$\tau_q = \text{Sat}_M\left(10\sin\hat{q} - \hat{\delta} + \ddot{q}_r - A_{0\sim1}\hat{s}_e([\hat{q}, \hat{\dot{q}}]^\mathsf{T} - [q_r, \dot{q}_r]^\mathsf{T})\right). \tag{5.114}$$

The other parameters are set as $A_{0\sim1} = [5, 2]^\mathsf{T}, r = 1.5, \epsilon = 4, M = 11.5, a_0 = 0.9, b_0 = 0.8$. The comparison experimental results of finite-time fault-tolerant tracking control and asymptotic fault-tolerant tracking control are shown in Figs. 5.2, 5.3, and 5.4.

From Fig. 5.2, it can be seen that the robot system (5.111) can track the given signal within a finite-time under the controller (5.114). Due to the suddenness of the fault and the sampling time interval, when the fault occurs, the tracking error under the existing FAS controller will be amplified, as shown in Figs. 5.2 and 5.3. From Fig. 5.4a, it can be known that the finite-time fault estimation has higher estimation accuracy and faster estimation speed. The early fluctuation phenomenon is normal because the state observation error and fault estimation error have not converged in the early stage. After the fault occurs, the fault compensation speed of the traditional

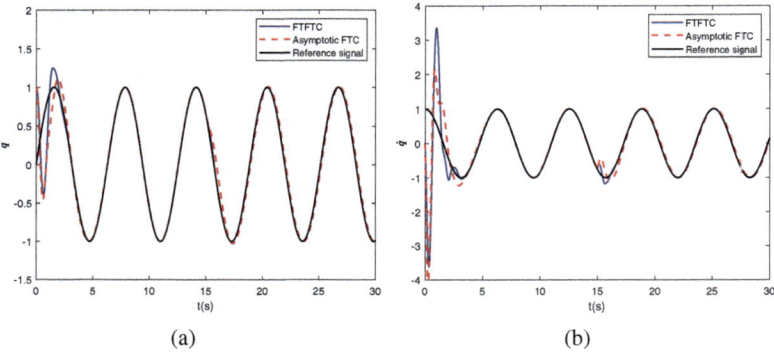

Fig. 5.2 The state trajectories in the FTC experiment

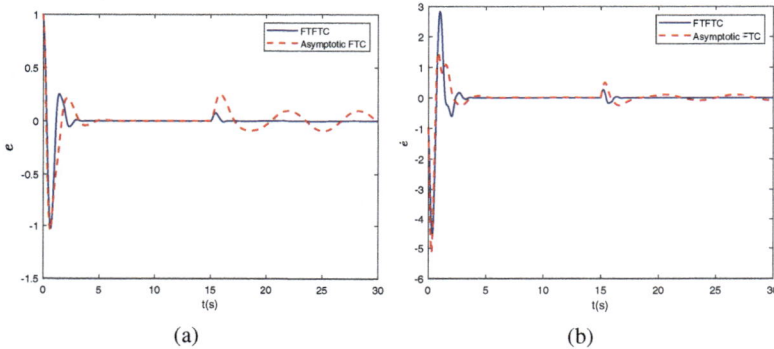

Fig. 5.3 The error trajectories in the FTC experiment

FTC is slow, while the fault compensation speed of the proposed FTFTC in this monograph is faster. The specific evolution curves of the two controllers are shown in Fig. 5.4b.

5.4.3 An FTC Experiment of a Rigid Aircraft Model

Consider the following a short-period dynamic rigid aircraft model [115]

$$
\begin{cases}
\dot{\alpha} = -\frac{\mathscr{L}_\alpha}{v_\alpha}\alpha + p, \\
\dot{p} = \mathscr{M}_\alpha\dot{\alpha} + \mathscr{M}_p p + \mathscr{M}_u(u + \mathfrak{f}(\alpha, u)), \\
y = \alpha + v,
\end{cases}
\tag{5.115}
$$

where $\alpha, p, u, \mathscr{L}_\alpha, v_\alpha, \mathscr{M}_\alpha, \mathscr{M}_p, \mathscr{M}_u$ are the angle of attack, pitch rate, lift moment, slope of the lift curve, corrected airspeed, pitch moment, pitch damping and

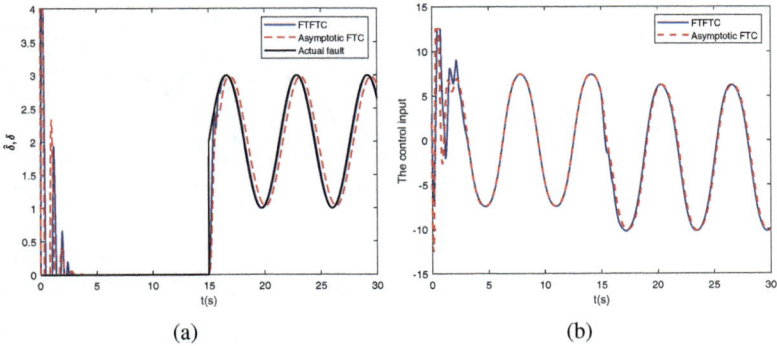

Fig. 5.4 The fault and input trajectories in the FTC experiment

lift moment coefficient, respectively. The measurement noise v follows a normal distribution, and the unknown nonlinear form $\mathfrak{f}(\alpha, u)$ is

$$\mathfrak{f}(\alpha, u) = \delta + [(1 - c_0)e^{-\frac{(\alpha - \alpha_0)^2}{\xi_0^2}} + c_0][\tanh(u + h_0) + \tanh(u - h_0) + 0.01u], \tag{5.116}$$

where α_0 represents the initial angle, δ represents the fault signal, and c_0, ξ_0, and h_0 are unknown constants. By defining $x = \alpha$, the aircraft model can be transformed into the following FAS model

$$\ddot{x} = F(x^{(0\sim1)}) + G(x^{(0\sim1)})u + \delta_f, \tag{5.117}$$

where

$$F(x^{(0\sim1)}) = (\mathcal{M}_\alpha + \mathcal{M}_p - \frac{\mathcal{L}_\alpha}{v_\alpha})\dot{x} + \frac{\mathcal{M}_p \mathcal{L}_\alpha}{v_\alpha}x, \tag{5.118}$$

$$G(x^{(0\sim1)}) = \mathcal{M}_u, \delta_f = \mathcal{M}_u \mathfrak{f}(x, u). \tag{5.119}$$

The trajectory tracking task is to have α track the reference signal $\alpha_r = \sin t$. In the finite-time fault-tolerant tracking control experiment design, the observer and controller are designed as (5.52) and (5.55). The experimental parameters are set as $\mathcal{L}_\alpha = -511.55$, $v_\alpha = 502$, $\mathcal{M}_\alpha = 0.82$, $\mathcal{M}_p = -1.08$, $\mathcal{M}_u = -0.18$, $\alpha_0 = 5$, $c_0 = 0.5$, $\xi_0 = 0.5$, $h_0 = 0.25$ [90], and the fault model is

$$\delta(t) = \begin{cases} 0, 0 \leq t < 15\text{s}, \\ 10 + 2\sin(t - 15), t \geq 15\text{s}. \end{cases} \tag{5.120}$$

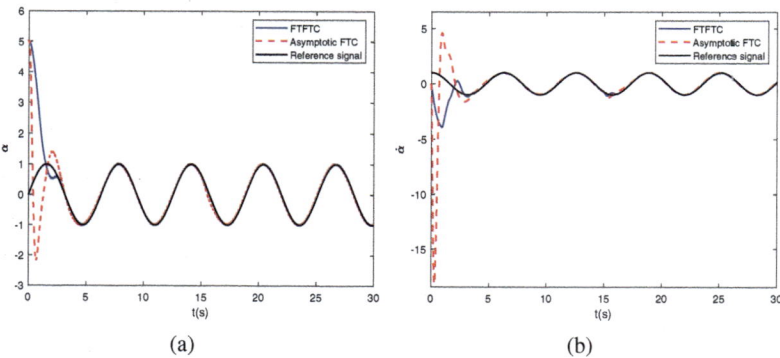

Fig. 5.5 The signal trajectories in the rigid aircraft experiment

The parameters are $A_{0\sim1} = [5, 2]^{\mathsf{T}}$, $B_0 = 2$, $B_1 = 4$, $B_2 = 3$, $r = 1.5$, $\epsilon = 4$, $M = 15$, $a_0 = 0.9$, $b_0 = 0.8$. The results are shown in Fig. 5.5. The FTFTC compensates for the fault more quickly at $t = 15$ s.

5.5   Notes and References

This chapter presents a finite-time parameterized controller structure for nonlinear FASs. In the problem of stabilizing control for measurable FASs, the finite-time controller has the advantages of a standard structure and clear parameter tuning rules. In the problem of fault-tolerant tracking control for unmeasurable FASs, the proposed FTFTC framework can quickly provide fault estimations and complete the trajectory tracking task. This FTFTC is an important supplement to the FAS theory in terms of control performance and system safety, and further expands its application space in rapid control systems.

Chapter 6
Self-Healing FTC for FASs Against Sensor Faults

The chapter proposes a novel self-healing fault accommodation framework for nonlinear fully actuated system (FASs) under faulty sensors. Beginning from a nonlinear FAS model with multiple measurements, a d-redundant observability proposition is derived from the observable structure under each sensor measurement. Since the error dynamics is ultimately uniformly bounded, a definition of ideal sensor fault tolerance is determined. After a necessary and sufficient fault tolerance condition is developed, a self-healing fault-tolerant control (SHFTC) framework is proposed, which can cope with sensor faults during steady-state or transient processes. The main results are proved theoretically and illustrated experimentally.

6.1 Overview of Self-Healing Fault-Tolerant Control Mechanisms

Unlike the fault-tolerant control (FTC) designs for component and actuator faults in previous chapters, the FTC design for sensor faults requires certain system redundancy. If all the measurement sensors fail, the general FTC scheme will not work normally. Most existing sensor FTC methods actually incorporate redundancy ideas, such as model structure redundancy [13, 114], hardware redundancy [19, 86], and distributed redundancy [79, 103, 135]. Although there are potential redundancy assumptions, these studies do not pay sufficient attention to the analysis of system redundancy. How to extract the redundant features of state equations and measurement equations is the essence of sensor FTC design. If sensor faults occur during the transient process, the faults cannot be analyzed from the steady-state perspective, but the state observation during the transient process will be helpful for sensor fault diagnosis and FTC research. On the other hand, the nonlinear cancellation principle in the FAS theory relies on precise system states, and unreliable sensor

© The Author(s) 2026
D. Zhou, M. Cai, *Fault-Tolerant Control for Fully Actuated Systems*,
https://doi.org/10.1007/978-981-95-0691-0_6

measurements will seriously disrupt the closed-loop linear structure of FASs. Therefore, it is necessary to study the sensor FTC technology for FASs.

Redundancy gives dynamic systems a certain degree of self-healing ability. The concept of self-healing mainly appears in the field of smart grids, indicating that the system can operate safely through technologies such as self-diagnosis, self-decision, and self-repair [3]. The SHFTC is a novel active fault-tolerant control (AFTC) technology for implementing corresponding strategies in the event of system faults, as it includes the two most important parts in the AFTC—fault diagnosis and controller switching, so the SHFTC belongs to the category of AFTC in a certain sense.

This chapter proposes an SHFTC framework for high-order FASs with sensor faults. A d-redundant observability condition is derived based on the observable form of each measurement. Based on the ultimately bounded error dynamics, a new definition of ideal sensor fault tolerance is given. The proposed SHFTC structure can be applied to the steady-state or transient processes of dynamic systems.

6.2 Problem Formulation and Preliminaries

Consider a class of FASs with sensor faults:

$$\begin{cases} x^{(n)} = F(x^{(0\sim n-1)}) + G(x^{(0\sim n-1)})u, \\ y_j = H_j(x^{(0\sim n-1)}) + s_j + v_j, \ j \in [m], \end{cases} \tag{6.1}$$

where $x \in \mathbb{R}$ is the basic state, $u \in \mathbb{R}$ is the control input, $y_j(t) \in \mathbb{R}$ is the measurement out, and $F(\cdot), G(\cdot), H_j(\cdot), j \in [m]$ are smooth nonlinear functions. The system state is within the compact set, i.e., $x^{(0\sim n-1)} \in \Omega_x \triangleq \{\varsigma \mid \varsigma \in \mathbb{R}^n : \|\varsigma\|_\infty \leq \varXi_x\}$ where $\varXi_x$ is a positive constant. The compactness requirement has become a prerequisite for many nonlinear system theories [5, 45, 55]. $v \triangleq [v_1, v_2, \cdots, v_m]^\mathsf{T}$ is a bounded measurement noise. $s_j, j \in [m]$ are possible sensor faults, and the fault signals may be unbounded in amplitude or non-differentiable. The FAS model (6.1) can describe many practical objects, such as the mechanical arm given by the Lagrangian equation [23]. From a mathematical perspective, the FAS model (6.1) can be derived from many systems, for example, hybrid-order generalized strict-feedback systems can be transformed into FASs [24].

Assumption 6.1 The functions $F(\cdot)$ and $G(\cdot)$ are Lipschitz, and their partial derivatives are bounded. The FAS model (6.1) satisfies the full-actuation condition, i.e., $\det G(x^{(0\sim n-1)}) \neq 0$ or $\infty, \forall x^{(0\sim n-1)} \in \Omega_x$.

The controller design of the FAS model (6.1) is challenging because the system state cannot be directly measured, and most classical output feedback will fail due to sensor faults. The FTC design without amplitude constraints for sensor faults requires redundant information, which includes hardware redundancy, data redundancy, channel redundancy, etc. Therefore, the purpose of this chapter is to analyze the redundant information of the FAS and design an FTC structure based on redundant feature extraction.

6.2.1 Controllability and Global Stability

In a sense, the FAS model (6.1) corresponds to the complete controllability of nonlinear systems and can yield the global stability of nonlinear systems.

Lemma 6.1 (Controllability [28]) *If a nonlinear system can be equivalently transformed into a global FAS model, then the nonlinear system is completely controllable.*

Assuming that the FAS model (6.1) is perfectly ideal, i.e., full-order states are measurable and there are no sensor faults in the system, the following FAS controller is designed as [23]

$$u = -G^{-1}(x^{(0 \sim n-1)}) \left[F(x^{(0 \sim n-1)}) + A_{0 \sim n-1} x^{(0 \sim n-1)} - u_a \right]. \tag{6.2}$$

Then the controller (6.2) will generate a pre-closed-loop linear system

$$\dot{x}^{(0 \sim n-1)} = \Phi_A x^{(0 \sim n-1)} + B_n u_a, \tag{6.3}$$

where u_a serves as an auxiliary signal for achieving trajectory tracking and other tasks,

$$\Phi_A = \begin{bmatrix} 0 & 1 & 0 & \cdots & 0 \\ 0 & 0 & 1 & \cdots & 0 \\ \vdots & & & \ddots & \vdots \\ 0 & 0 & 0 & \cdots & 1 \\ -A_0 & -A_1 & -A_2 & \cdots & -A_{n-1} \end{bmatrix} \in \mathbb{R}^{n \times n}, \quad B_n = \begin{bmatrix} 0 \\ 0 \\ \vdots \\ 0 \\ 1 \end{bmatrix} \in \mathbb{R}^n. \tag{6.4}$$

When there exists a positive definite matrix $P \in \mathbb{R}^{n \times n}$ that is a solution to the equation $\Phi_A^T P + P \Phi_A = -\kappa_1 I_n$, where κ_1 is a normal constant, the FAS controller (6.2) can yield the global stability of the nonlinear system. However, most FASs often encounter nonideal situations such as faults and noise. In such cases, the system state cannot be directly measured, and thus (6.2) and (6.3) cannot be directly obtained. Therefore, it is necessary to study the design of fault-tolerant observers and controllers for FASs under sensor faults.

6.2.2 Observability and Redundant Observability

Considering that the FAS model (6.1) has m sensors for measurement, the partial state information provided by each individual measurement is of great significance. For each measurement y_j, the relative order is r_j, i.e., for $\forall t \geq 0$, it turns out that

$$\mathscr{L}_\mathscr{G} \mathscr{L}_\mathscr{F}^{r_j-1} H_j(x^{(0 \sim n-1)}) \neq 0, \mathscr{L}_\mathscr{G} \mathscr{L}_\mathscr{F}^{k} H_j(x^{(0 \sim n-1)}) = 0, k \in [r_j - 2], \quad (6.5)$$

where

$$\begin{cases} \mathscr{F} = [\dot{z}, \ddot{z}, \cdots, z^{(n-1)}, F(x^{(0 \sim n-1)})]^\mathsf{T}, \\ \mathscr{G} = [0, 0, \cdots, 0, G(x^{(0 \sim n-1)})]^\mathsf{T}. \end{cases} \quad (6.6)$$

Define a differential diffeomorphism $\hbar_j : \mathbb{R}^n \to \mathbb{R}^{r_j} \times \mathbb{R}^{n-r_j}$, $[\xi_j^\mathsf{T}, \eta_j^\mathsf{T}]^\mathsf{T} = \hbar_j(x^{(0 \sim n-1)})$,

$$\xi_j = \begin{bmatrix} \xi_{j,1} \\ \xi_{j,2} \\ \vdots \\ \xi_{j,r_j} \end{bmatrix} = \begin{bmatrix} \mathscr{L}_\mathscr{F} H_j(x^{(0 \sim n-1)}) \\ \mathscr{L}_\mathscr{F}^2 H_j(x^{(0 \sim n-1)}) \\ \vdots \\ \mathscr{L}_\mathscr{F}^{r_j-1} H_j(x^{(0 \sim n-1)}) \end{bmatrix} \triangleq \Upsilon_j(x^{(0 \sim n-1)}) \in \mathbb{R}^{r_j}, \quad (6.7)$$

the FAS under each measurement is [39]

$$\dot{\xi}_j = \begin{bmatrix} \dot{\xi}_{j,1} \\ \dot{\xi}_{j,2} \\ \vdots \\ \dot{\xi}_{j,r_j} \end{bmatrix} = \begin{bmatrix} \xi_{j,2} \\ \vdots \\ \xi_{j,r_j} \\ F_o^j(\xi_j) \end{bmatrix} + \begin{bmatrix} 0 \\ \vdots \\ 0 \\ G_o^j(\xi_j) \end{bmatrix} u, \quad (6.8a)$$

$$\dot{\eta}_j = F_{uo}^j(\xi_j, \eta_j) + G_{uo}^j(\xi_j, \eta_j)u, \quad (6.8b)$$

$$y_j = \xi_{j,1} + s_j + v_j, \quad (6.8c)$$

where $F_o^j(\cdot)$, $G_o^j(\cdot)$, $F_{uo}^j(\cdot)$, $G_{uo}^j(\cdot)$ are the transformed functions, and the states are in compact sets, i.e., $\xi_j \in \Omega_j \triangleq \{\varsigma \mid \varsigma \in \mathbb{R}^{r_j} : \|\varsigma\|_\infty \leq \varXi_j\}$ with $\varXi_j > 0$, $j \in [m]$.

Lemma 6.2 (Observerability [39]) *If a nonlinear system can be equivalently transformed into the model (6.8a), then this nonlinear system is completely observable.*

For each measurement y_j, (6.8) is the observable structure decomposition of the FAS model (6.1), where ξ_j represents the observable subspace and η_j represents the unobservable subspace. The fused observable space represents the entire observable structure of the FAS model (6.1), i.e.,

$$
\xi = \begin{bmatrix} \xi_1 \\ \xi_2 \\ \vdots \\ \xi_m \end{bmatrix} = \begin{bmatrix} \Upsilon_1(x^{(0\sim n-1)}) \\ \Upsilon_2(x^{(0\sim n-1)}) \\ \vdots \\ \Upsilon_m(x^{(0\sim n-1)}) \end{bmatrix} \triangleq \Upsilon_{[m]}(x^{(0\sim n-1)}) \in \mathbb{R}^{\sum_{j\in[m]} r_j}. \tag{6.9}
$$

Definition 6.1 (Injective Immersion Mapping [41, 102]) If the Jacobian matrix corresponding to the mapping $\Upsilon_{[m]} : \Omega_x \to \Upsilon_{[m]}(\Omega_x)$ is of full column rank, then $\Upsilon_{[m]}$ is called an injective immersion mapping.

Proposition 6.1 (Observability) *If the mapping $\Upsilon_{[m]} : \Omega_x \to \Upsilon_{[m]}(\Omega_x)$ is an injective immersion mapping, then the nonlinear FAS (6.1) is completely observable.*

Proof According to Lemma 6.2, ξ_j, $j \in [m]$ correspond to m observable subspaces. The complete observability of the FAS (6.1) depends on whether ξ can recover $x^{(0\sim n-1)}$ through the inversion operation. When the mapping $\Upsilon_{[m]}$ is an injective immersion mapping on the compact set Ω_x, this mapping is bi-Lipschitz [55] and has a left inverse mapping $\Upsilon_{[m]}^{-1}$.

Remark 6.1 Unlike the observability condition [22], Proposition 6.1 analyzes the observability condition of FASs from another perspective, and it is the information fusion of state observations.

In order to conduct redundancy analysis for multiple sensors, a sensor set indexed by $I, I \subset [m]$ is established. The cardinality of this set is denoted as $|I|$. The observable structure under the sensor set I is

$$
\xi_I = \begin{bmatrix} \xi_{I_1} \\ \xi_{I_2} \\ \vdots \\ \xi_{I_{|I|}} \end{bmatrix} = \begin{bmatrix} \Upsilon_{I_1}(x^{(0\sim n-1)}) \\ \Upsilon_{I_2}(x^{(0\sim n-1)}) \\ \vdots \\ \Upsilon_{I_{|I|}}(x^{(0\sim n-1)}) \end{bmatrix} \triangleq \Upsilon_I(x^{(0\sim n-1)}) \in \mathbb{R}^{\sum_{j\in I} r_j}. \tag{6.10}
$$

Proposition 6.2 (Redundant Observability [55]) *If for any subset $I \subset [m]$ with $|I| = m - d$, the mapping $\Upsilon_I : \Omega_x \to \Upsilon_I(\Omega_x)$ is an injective immersion mapping, then the FAS (6.1) is d-redundant observable. And the 0-redundant observability corresponds to the traditional observability in Proposition 6.1.*

Proof The proof process is similar to Proposition 6.1.

According to the injective immersion mapping condition, the bi-Lipschitz mappings Υ and Υ_I both have corresponding left inverse mappings. However, as shown in Fig. 6.1, some observed values deviate from the original domain of the actual state. To ensure the consistency of the mapping domain, the Lipschitz extension

Fig. 6.1 The signal
trajectories of $\xi_j, \hat{\xi}_j$

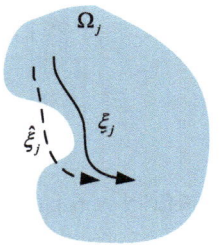

technique [88] is utilized to define the two modified left inverse mappings as

$$\text{INV}(\Upsilon_{[m]})(\xi) : \mathbb{R}^{\sum_{j\in[m]} r_j} \to \Omega_x$$

$$\xi \mapsto \text{Sat}_{\varXi_x}\left(\Upsilon_{[m]}^{-1}\left([\text{Sat}_{\varXi_1}(\xi_1), \text{Sat}_{\varXi_2}(\xi_2), \cdots, \text{Sat}_{\varXi_m}(\xi_m)]^{\mathsf{T}}\right)\right),$$

$$(6.11)$$

$$\text{INV}(\Upsilon_I)(\xi_I) : \mathbb{R}^{\sum_{j\in I} r_j} \to \Omega_x$$

$$\xi_I \mapsto \text{Sat}_{\varXi_x}\left(\Upsilon_I^{-1}\left([\text{Sat}_{\varXi_{I_1}}(\xi_{I_1}), \text{Sat}_{\varXi_{I_2}}(\xi_{I_2}), \cdots, \text{Sat}_{\varXi_{I_{|I|}}}(\xi_{I_{|I|}})]^{\mathsf{T}}\right)\right),$$

$$(6.12)$$

where $\Upsilon_{[m]}^{-1}, \Upsilon_I^{-1}$ are the left inverse mappings of $\Upsilon_{[m]}, \Upsilon_I$, respectively, and $\text{Sat}_{\varXi_x}(\cdot) = \min\{\max\{\cdot, -\varXi_x\}, \varXi_x\}$ is a saturation function. The definitions of the other Sat functions are similar.

6.3 Design Without Sensor Faults: Fusion Observation and Control

The control task of the original FAS (6.1) is to make x track the smooth reference signal x_r, where the reference signal satisfies $\sup_{t\geq 0} \|x_r^{(0\sim n)}\| < \infty$. Proposition 6.1 is a sufficient condition for the fusion observation and controller design of the original FAS (6.1). For a fault-free FAS, let $X = [x_1, x_2, \cdots, x_n]^{\mathsf{T}} \triangleq x^{(0\sim n-1)}$, and the integrated structure of the fusion observer and controller is designed as:

$$\begin{cases} \dot{\hat{\xi}}_j = \begin{bmatrix} \hat{\xi}_{j,2} \\ \vdots \\ \hat{\xi}_{j,r_j} \\ F_o^j(\hat{\xi}_j) \end{bmatrix} + \begin{bmatrix} \ell L_{j,1} \\ \ell^2 L_{j,2} \\ \vdots \\ \ell^{r_j} L_{j,r_j} \end{bmatrix} (y_j - \hat{\xi}_{j,1}) + \begin{bmatrix} 0 \\ \vdots \\ 0 \\ G_o^j(\hat{\xi}_j) \end{bmatrix} u, \; j \in [m], \\ \hat{X} = \text{INV}(\Upsilon_{[m]})(\hat{\xi}), \; \hat{\xi} = [\hat{\xi}_1^{\mathsf{T}}, \hat{\xi}_2^{\mathsf{T}}, \cdots, \hat{\xi}_m^{\mathsf{T}}]^{\mathsf{T}}, \\ u = -G^{-1}(\hat{X})\left[F(\hat{X}) - x_r^{(n)} + A_{0\sim n-1}\left(\hat{X} - x_r^{(0\sim n-1)}\right)\right], \end{cases}$$

$$(6.13)$$

where $\hat{X}, \hat{\xi}$ satisfy $\hat{\xi}_j(0) \in \Omega_j$, $j \in [m]$, ℓ is a positive constant greater than 1, and $L_{j,1}, L_{j,2}, \cdots, L_{j,r_j}$, $j \in [m]$ are the observer parameters.

Theorem 6.1 *For the fault-free observable FAS* (6.1) *under the integrated framework of observer and controller* (6.13), *there exist positive constants* t_1, t_2, M_1, M_2, M_3 *such that the estimation error and the tracking error are ultimately uniformly bounded, i.e.,*

$$
\begin{cases}
\|\xi_{j,k} - \hat{\xi}_{j,k}\| \le \ell^{k-r_j-1} M_1, \ j \in [m], k \in [r_j], \forall t > t_1, \\
\|x^{(0 \sim n-1)} - \hat{X}\| \le M_2/\ell, \forall t > t_1, \\
\|x - x_r\| \le M_3/\ell, \forall t > t_2,
\end{cases}
\tag{6.14}
$$

Furthermore, the tracking error converges to a neighborhood near the origin. The parameter $\ell > 1$ can result in a tiny tracking error.

Proof Define the observation error of the jth observable space

$$
\varepsilon_{j,k} = \ell^{r_j-k}(\xi_{j,k} - \hat{\xi}_{j,k}),
\tag{6.15}
$$

and the derivative of $\varepsilon_j = [\varepsilon_{j,1}, \varepsilon_{j,2}, \cdots, \varepsilon_{j,r_j}]^{\mathsf{T}}$ is

$$
\begin{cases}
\dot{\varepsilon}_{j,k} = \ell \varepsilon_{j,k+1} - \ell L_{j,k} \varepsilon_{j,1} - \ell^{r_j} L_{j,k} v_j, k \in [r_j - 1], \\
\dot{\varepsilon}_{j,r_j} = \Theta_{1j} - \ell L_{j,r_j} \varepsilon_{j,1} - \ell^{r_j} L_{j,r_j} v_j,
\end{cases}
\tag{6.16}
$$

where

$$
\Theta_{1j} = F_o^j(\xi_j) - F_o^j(\hat{\xi}_j) + [G_o^j(\xi_j) - G_o^j(\hat{\xi}_j)]u.
\tag{6.17}
$$

The actual observation error of the FAS model (6.1) is

$$
\tilde{X} = x^{(0 \sim n-1)} - \hat{X} = \text{INV}(\Upsilon_{[m]})(\xi) - \text{INV}(\Upsilon_{[m]})(\hat{\xi}).
\tag{6.18}
$$

Based on the Lipschitz property of the mapping $\text{INV}(\Upsilon_{[m]})$, it follows that

$$
\|\tilde{X}\| \le \overline{\text{Lip}}(\text{INV}(\Upsilon_{[m]})) \|\xi - \hat{\xi}\| \le \overline{\text{Lip}}(\text{INV}(\Upsilon_{[m]})) \|\varepsilon\|, \varepsilon = [\varepsilon_1^{\mathsf{T}}, \varepsilon_2^{\mathsf{T}}, \cdots, \varepsilon_m^{\mathsf{T}}]^{\mathsf{T}}.
\tag{6.19}
$$

Define the high-order tracking error

$$
E = [e_1, e_2, \cdots, e_n]^{\mathsf{T}} = e^{(0 \sim n-1)} = x^{(0 \sim n-1)} - x_r^{(0 \sim n-1)},
\tag{6.20}
$$

and the derivative of E is

$$
\begin{cases}
\dot{e}_i = e_{i+1}, i \in [n-1], \\
\dot{e}_n = \Theta_2 + A_{0 \sim n-1}(x^{(0 \sim n-1)} - \hat{X}) - A_{0 \sim n-1}(x^{(0 \sim n-1)} - x_r^{(0 \sim n-1)}),
\end{cases}
\tag{6.21}
$$

where

$$\Theta_2 = F(x^{(0\sim n-1)}) - F(\hat{X}) + [G(x^{(0\sim n-1)}) - G(\hat{X})]u. \tag{6.22}$$

Thus the overall error dynamics is

$$\begin{cases} \dot{\varepsilon}_{j,k} = \ell\varepsilon_{j,k+1} - \ell L_{j,k}\varepsilon_{j,1} - \ell^{r_j}L_{j,k}v_j, \\ \dot{\varepsilon}_{j,r_j} = \Theta_{1j} - \ell L_{j,r_j}\varepsilon_{j,1} - \ell^{r_j}L_{j,r_j}v_j, \\ \dot{e}_i = e_{i+1}, \\ \dot{e}_n = \Theta_2 + A_{0\sim n-1}\tilde{X} - A_{0\sim n-1}E, \end{cases} \tag{6.23}$$

where $j \in [m], k \in [r_j - 1], i \in [n - 1]$.

A Lyapunov function is chosen as

$$V = \sum_{j=1}^{m} \varepsilon_j^{\mathsf{T}} \mathscr{P}_j \varepsilon_j + E^{\mathsf{T}} P E. \tag{6.24}$$

$\mathscr{P}_j, j \in [m]$ are the positive definite solutions to the following Lyapunov equations

$$\mathscr{P}_j \Psi_j + \Psi_j^{\mathsf{T}} \mathscr{P}_j = -\kappa_2 I_n, j \in [m], \tag{6.25}$$

where

$$\Psi_j = \begin{bmatrix} -L_{j,1} & 1 & 0 & \cdots & 0 \\ -L_{j,2} & 0 & 1 & \cdots & 0 \\ \vdots & & & \ddots & \\ -L_{j,r_j-1} & 0 & 0 & \cdots & 1 \\ -L_{j,r_j} & 0 & 0 & \cdots & 0 \end{bmatrix} \in \mathbb{R}^{r_j \times r_j}, \tag{6.26}$$

κ_2 is a designed parameter, and $\Psi_j, j \in [m]$ are Hurwitz matrices. The derivative of V is

$$\frac{dV}{dt} = \sum_{j=1}^{m} \left\{ \ell\varepsilon_j^{\mathsf{T}}(\mathscr{P}_j\Psi_j + \Psi_j^{\mathsf{T}}\mathscr{P}_j)\varepsilon_j - 2\ell^{r_j}v_j L_j\mathscr{P}_j\varepsilon_j + 2\sum_{j=1}^{m}\Theta_{1j}B_{r_j}^{\mathsf{T}}\mathscr{P}_j\varepsilon_j \right\}$$

$$+ E^{\mathsf{T}}(P\Phi_A + \Phi_A^{\mathsf{T}}P)E + 2\Theta_2 B_n^{\mathsf{T}}PE + 2A_{0\sim n-1}\tilde{X}B_n^{\mathsf{T}}PE$$

$$= -\kappa_2 \sum_{j=1}^{m} \ell\varepsilon_j^{\mathsf{T}}\varepsilon_j - \kappa_1 E^{\mathsf{T}}E + \sum_{j=1}^{m} \left\{ 2\Theta_{1j}B_{r_j}^{\mathsf{T}}\mathscr{P}_j\varepsilon_j - 2\ell^{r_j}v_j L_j\mathscr{P}_j\varepsilon_j \right\}$$

$$+ 2\Theta_2 B_n^{\mathsf{T}}PE + 2A_{0\sim n-1}\tilde{X}B_n^{\mathsf{T}}PE, \tag{6.27}$$

where $L_j = [L_{j,1}, L_{j,2}, \cdots, L_{j,r_j}]$, $B_{r_j} = [0, 0, \cdots, 0, 1]^{\mathsf{T}} \in \mathbb{R}^{r_j}$, $j \in [m]$.

Based on the Lipschitz properties of F_o^j and G_o^j, Assumption 6.1, the bounded-ness of the high-order derivatives of x_r, the boundedness of the mapping $\mathrm{INV}(\Upsilon_{[m]})$, $\|\xi_j - \hat{\xi}_j\| \le \|\varepsilon_j\|$ and $\|\tilde{X}\| \le \overline{\mathrm{Lip}}(\mathrm{INV}(\Upsilon_{[m]}))\|\varepsilon\|$, the analysis of Θ_{1j} yields

$$
\begin{aligned}
\Theta_{1j} &= F_o^j(\xi_j) - F_o^j(\hat{\xi}_j) + \left[G_o^j(\xi_j) - G_o^j(\hat{\xi}_j)\right]u \\
&\le a_1 \|\varepsilon_j\| + a_2 \left[a_3 + \|A_{0\sim n-1}\|(\|\tilde{X}\| + \|E\|)\right] \\
&\le a_4 \|\varepsilon\| + a_5 \|E\| + a_6,
\end{aligned}
\tag{6.28}
$$

where $a_1, a_2, a_3, a_4, a_5, a_6$ are positive constants. Similarly,

$$
\begin{aligned}
\Theta_2 &= F(x^{(0\sim n-1)}) - F(\hat{X}) + [G(x^{(0\sim n-1)}) - G(\hat{X})]u \\
&\le a_7 \|\tilde{X}\| + a_8 \|E\| + a_9 \\
&\le a_{10} \|\varepsilon\| + a_8 \|E\| + a_9,
\end{aligned}
\tag{6.29}
$$

where a_7, a_8, a_9, a_{10} are positive constants. By combining (6.27)–(6.29), it can be obtained that

$$
\begin{aligned}
\frac{\mathrm{d}V}{\mathrm{d}t} &< -\kappa_2 \ell \|\varepsilon\|^2 - \kappa_1 \|E\|^2 + mb_1 \ell^n \|v\|_\infty \|\varepsilon\| + mb_2 \|\varepsilon\|(a_4 \|\varepsilon\| + a_5 \|E\| + a_6) \\
&\quad + b_3 \|E\|(a_{10} \|\varepsilon\| + a_8 \|E\| + a_9) + b_4 \|\varepsilon\| \|E\| \\
&\le -(\kappa_2 \ell - ma_4 b_2)\|\varepsilon\|^2 + (mb_1 \ell^n \|v\|_\infty + ma_6 b_2)\|\varepsilon\| \\
&\quad - (\kappa_1 - a_8 b_3)\|E\|^2 + a_9 b_3 \|E\| + (ma_5 b_2 + a_{10} b_3 + b_4)\|\varepsilon\| \|E\| \\
&\le -\left(\frac{\kappa_2 \ell}{2} - ma_4 b_2\right)\|\varepsilon\|^2 + (mb_1 \ell^n \|v\|_\infty + ma_6 b_2)\|\varepsilon\| \\
&\quad - \left(\kappa_1 - a_8 b_3 - \frac{(ma_5 b_2 + b_5)^2}{2\kappa_2 \ell}\right)\|E\|^2 + a_9 b_3 \|E\|,
\end{aligned}
\tag{6.30}
$$

where $b_1 = 2 \max_{j \in [m]} \|L_j \mathscr{P}_j\|$, $b_2 = 2 \max_{j \in [m]} \|B_{r_j}^{\mathsf{T}} \mathscr{P}_j\|$, $b_3 = 2\|B_n^{\mathsf{T}} P\|$,

$$
b_4 = 2\overline{\mathrm{Lip}}(\mathrm{INV}(\Upsilon_{[m]}))\|A_{0\sim n-1}\|\|B_n^{\mathsf{T}} P\|,
$$

and $b_5 = a_{10} b_3 + b_4$. If the parameters meet the following condition

$$
\ell > \max\left\{1, \frac{2ma_4 b_2}{\kappa_2}, \frac{(ma_5 b_2 + b_5)^2}{2\kappa_2(\kappa_1 - a_8 b_3)}\right\},
\tag{6.31}
$$

then the observation error and the tracking error are ultimately uniformly bounded, i.e., there exists $t_0 > 0$ such that for $t > t_0$, the observation error and the tracking error are bounded. The ultimate error is mainly influenced by the noise υ and the system order n, and the main influencing term is $\ell^n \|\upsilon\|_\infty$. The lower the system order, the smaller the error bound. When the system is not contaminated by noise, the error bound will significantly decrease. After applying the bounded overall error to (6.28), the analysis of $V_j = \varepsilon_j^\mathsf{T} \mathscr{P}_j \varepsilon_j, t > t_0, j \in [m]$ yields

$$\frac{dV_j}{dt} < -\kappa_2 \ell \|\varepsilon_j\|^2 + c_1 \|\varepsilon_j\| \leq -\frac{\kappa_2 \ell V_j}{\lambda_{\max}(\mathscr{P}_j)} + \frac{c_1 \sqrt{V_j}}{\sqrt{\lambda_{\min}(\mathscr{P}_j)}}, \tag{6.32}$$

where c_1 is a positive constant. For $\forall t > t_0$, it follows that

$$\sqrt{V_j(\varepsilon_j(t))} < \sqrt{V_j(\varepsilon_j(t_0))} \exp(-c_2 \ell(t - t_0)) + c_3 \int_{t_0}^{t} \exp(-c_2 \ell(t - \tau)) d\tau, \tag{6.33}$$

where $c_2 = 0.5\kappa_2 / \lambda_{\max}(\mathscr{P}_j)$, $c_3 = 0.5c_1 / \sqrt{\lambda_{\min}(\mathscr{P}_j)}$. Obviously, the observation error converges exponentially, i.e.,

$$\|\varepsilon_j\| \leq \max \left\{ \beta_j \exp(-\alpha_j \ell t), \mathscr{M}_j / \ell \right\}, j \in [m], \tag{6.34}$$

where $\alpha_j, \beta_j, \mathscr{M}_j$ are positive constants. For $j \in [m], k \in [r_j]$, there exists $t_1 > t_0$ such that

$$\|\xi_{j,k} - \hat{\xi}_{j,k}\| = \ell^{k-r_j} \|\varepsilon_{j,k}\| \leq \ell^{k-r_j-1} M_1, \forall t > t_1, \tag{6.35}$$

where M_1 is a positive constant. Therefore, the actual observational error satisfies $\|\tilde{X}\| \leq M_2 / \ell$ for all $t > t_1$, where M_2 is a positive constant.

Applying $\|\tilde{X}\| \leq M_2 / \ell, \forall t > t_1$ to (6.29), and analyzing $V_E = E^\mathsf{T} P E, t > t_1$ yields

$$\frac{dV_E}{dt} < -(\kappa_1 - c_4)\|E\|^2 + \frac{c_5}{\ell}\|E\|$$
$$\leq -\frac{(\kappa_1 - c_4) V_E}{\lambda_{\max}(P)} + \frac{c_5 \sqrt{V_E}}{\ell \sqrt{\lambda_{\min}(P)}}, \tag{6.36}$$

where c_4, c_5 are positive constants. For $\forall t > t_1$, it turns out that

$$\sqrt{V_E(E(t))} < \sqrt{V_E(E(t_1))} \exp(-c_6(t-t_1)) + \frac{c_7}{\ell} \int_{t_1}^{t} \exp(-c_6(t-\tau)) d\tau, \tag{6.37}$$

where $c_6 = 0.5(\kappa_1 - c_4)/\lambda_{\max}(P)$ and $c_7 = 0.5c_5 / \sqrt{\lambda_{\min}(P)}$. If $\kappa_1 > c_4$, then the tracking error converges, resulting in $\|E\| \leq M_3 / \ell$ for all $t > t_2$, where $t_2 > t_1$ and M_3 is a normal constant. Theorem 6.1 has been proved.

Corollary 6.1 *For the following class of fault-free, noise-free, observable FASs*

$$\begin{cases} x^{(n)} = F(x^{(0 \sim n-1)}) + gu, \\ y_j = C_j^\mathsf{T} x^{(0 \sim n-1)}, \ j \in [m], \end{cases} \tag{6.38}$$

where g is a non-zero constant, and $C_j \in \mathbb{R}^n$, $j \in [m]$ are constant vectors. The proposed integrated framework (6.13) can ensure that the observation error and tracking error are asymptotically stable.

Proof For the observable FAS (6.38), the observable and unobservable spaces under each measurement are

$$\dot{\xi}_j = \begin{bmatrix} \dot{\xi}_{j,1} \\ \dot{\xi}_{j,2} \\ \vdots \\ \dot{\xi}_{j,r_j} \end{bmatrix} = \begin{bmatrix} \xi_{j,2} \\ \vdots \\ \xi_{j,r_j} \\ F_o^j(\xi_j) \end{bmatrix} + \begin{bmatrix} 0 \\ \vdots \\ 0 \\ g_o^j \end{bmatrix} u, \tag{6.39a}$$

$$\dot{\eta}_j = F_{uo}^j(\xi_j, \eta_j) + g_{uo}^j u, \tag{6.39b}$$

$$y_j = \xi_{j,1}, \tag{6.39c}$$

where $F_o^j(\cdot)$, $F_{uo}^j(\cdot)$ is Lipschitz nonlinear functions, and g_o^j, g_{uo}^j, $j \in [m]$ are non-zero constants. The integrated framework for the fusion observer and controller is designed as follows

$$\begin{cases} \dot{\hat{\xi}}_j = \begin{bmatrix} \hat{\xi}_{j,2} \\ \vdots \\ \hat{\xi}_{j,r_j} \\ F_o^j(\hat{\xi}_j) \end{bmatrix} + \begin{bmatrix} \ell_j L_{j,1} \\ \ell_j^2 L_{j,2} \\ \vdots \\ \ell_j^{r_j} L_{j,r_j} \end{bmatrix} (y_j - \hat{\xi}_{j,1}) + \begin{bmatrix} 0 \\ \vdots \\ 0 \\ g_o^j \end{bmatrix} u, \ j \in [m], \\ \hat{X} = \mathrm{INV}(\Upsilon_{[m]})(\hat{\xi}), \ \hat{\xi} = [\hat{\xi}_1^\mathsf{T}, \hat{\xi}_2^\mathsf{T}, \cdots, \hat{\xi}_m^\mathsf{T}]^\mathsf{T}, \\ u = -g^{-1} \left[F(\hat{X}) - x_r^{(n)} + A_{0 \sim n-1} \left(\hat{X} - x_r^{(0 \sim n-1)} \right) \right], \end{cases} \tag{6.40}$$

Similar to the proof of Theorem 6.1, the observation error of the jth observable space is

$$\varepsilon_{j,k} = \ell^{r_j - k} (\xi_{j,k} - \hat{\xi}_{j,k}). \tag{6.41}$$

The derivative of $\varepsilon_j = [\varepsilon_{j,1}, \varepsilon_{j,2}, \cdots, \varepsilon_{j,r_j}]^\mathsf{T}$ is

$$\begin{cases} \dot{\varepsilon}_{j,k} = \ell \varepsilon_{j,k+1} - \ell L_{j,k} \varepsilon_{j,1}, \ k \in [r_j - 1], \\ \dot{\varepsilon}_{j,r_j} = \Theta_{1j} - \ell L_{j,r_j} \varepsilon_{j,1}, \end{cases} \tag{6.42}$$

where $\Theta_{1j} = F_o^j(\xi_j) - F_o^j(\hat{\xi}_j)$. The actual observation error of the FAS is

$$\tilde{X} = x^{(0\sim n-1)} - \hat{z}^{(0\sim n-1)} = \mathscr{T}(X) - \mathscr{T}(\hat{X}). \tag{6.43}$$

Based on the bi-Lipschitz property of the mapping $\mathrm{INV}(\Upsilon_{[m]})$, it can be obtained that

$$\|\tilde{X}\| \leq \overline{\mathrm{Lip}}(\mathrm{INV}(\Upsilon_{[m]}))\|\xi - \hat{\xi}\| \leq \overline{\mathrm{Lip}}(\mathrm{INV}(\Upsilon_{[m]}))\|\varepsilon\|, \varepsilon = [\varepsilon_1^\mathsf{T}, \varepsilon_2^\mathsf{T}, \cdots, \varepsilon_m^\mathsf{T}]^\mathsf{T}. \tag{6.44}$$

The dynamics of tracking error is

$$\begin{cases} \dot{e}_i = e_{i+1}, i \in [n-1], \\ \dot{e}_n = \Theta_2 + A_{0\sim n-1}(x^{(0\sim n-1)} - \hat{X}) - A_{0\sim n-1}(x^{(0\sim n-1)} - x_r^{(0\sim n-1)}), \end{cases} \tag{6.45}$$

where $\Theta_2 = F(x^{(0\sim n-1)}) - F(\hat{X})$. Thus, the overall error dynamics is

$$\begin{cases} \dot{\varepsilon}_{j,k} = \ell\varepsilon_{j,k+1} - \ell L_{j,k}\varepsilon_{j,1}, \\ \dot{\varepsilon}_{j,r_j} = \Theta_{1j} - \ell L_{j,r_j}\varepsilon_{j,1}, \\ \dot{e}_i = e_{i+1}, \\ \dot{e}_n = \Theta_2 + A_{0\sim n-1}\tilde{X} - A_{0\sim n-1}E, \end{cases} \tag{6.46}$$

where $j \in [m], k \in [r_j - 1], i \in [n-1]$.

A Lyapunov function is chosen as

$$V = \sum_{j=1}^{m} \varepsilon_j^\mathsf{T} \mathscr{P}_j \varepsilon_j + E^\mathsf{T} P E. \tag{6.47}$$

The derivative of V is

$$\frac{dV}{dt} = \sum_{j=1}^{m}\{\varepsilon_j^\mathsf{T} \mathscr{P}_j \dot{\varepsilon}_j + \dot{\varepsilon}_j^\mathsf{T} \mathscr{P}_j \varepsilon_j\} + E^\mathsf{T} P\dot{E} + \dot{E}^\mathsf{T} P E$$

$$= \sum_{j=1}^{m}\{\ell\varepsilon_j^\mathsf{T}(\mathscr{P}_j\Psi_j + \Psi_j^\mathsf{T}\mathscr{P}_j)\varepsilon_j + 2\sum_{j=1}^{m}\Theta_{1j}B_{r_j}^\mathsf{T}\mathscr{P}_j\varepsilon_j\}$$

$$+ E^\mathsf{T}(P\Phi_A + \Phi_A^\mathsf{T}P)E + 2\Theta_2(t)B_n^\mathsf{T}PE + 2A_{0\sim n-1}\tilde{X}B_n^\mathsf{T}PE$$

$$= -\kappa_2\sum_{j=1}^{m}\ell\varepsilon_j^\mathsf{T}\varepsilon_j - \kappa_1 E^\mathsf{T}E + 2\Theta_2 B_n^\mathsf{T}PE$$

$$+ \sum_{j=1}^{m} 2\Theta_{1j} B_{r_j}^{\mathsf{T}} \mathscr{P}_j \varepsilon_j + 2A_{0 \sim n-1} \tilde{X} B_n^{\mathsf{T}} P E. \tag{6.48}$$

Based on the Lipschitz properties of F_o^j and G_o^j, Assumption 6.1, the boundedness of the high-order derivatives of x_r, the boundedness of the mapping $\mathrm{INV}(\Upsilon_{[m]})$, $\|\xi_j - \hat{\xi}_j\| \le \|\varepsilon_j\|$ and $\|\tilde{X}\| \le \overline{\mathrm{Lip}}(\mathrm{INV}(\Upsilon_{[m]}))\|\varepsilon\|$, the analysis of Θ_{1j} and Θ_2 yields

$$\Theta_{1j} = F_o^j(\xi_j) - F_o^j(\hat{\xi}_j) \le a_{11}\|\varepsilon\|, \ \Theta_2 = F(x^{(0 \sim n-1)}) - F(\hat{X}) \le a_{12}\|\varepsilon\|, \tag{6.49}$$

where a_{11}, a_{12} are positive constants. By combining (6.48)–(6.49), it can be obtained that

$$\frac{\mathrm{d}V}{\mathrm{d}t} < -\kappa_2 \ell \|\varepsilon\|^2 - \kappa_1 \|E\|^2 + (b_6 + b_8)\|\varepsilon\|\|E\| + mb_7 \|\varepsilon\|^2$$

$$\le -\left(\frac{\kappa_2 \ell}{2} - mb_7\right)\|\varepsilon\|^2 - \left[\kappa_1 - \frac{(b_6 + b_8)^2}{2\kappa_2 \ell}\right]\|E\|^2, \tag{6.50}$$

that is, the derivative of the Lyapunov function can be completely negative definite, with $b_6 = 2a_{12}\|B_n^{\mathsf{T}} P\|$, $b_7 = 2a_{11} \max_{j \in [m]} \|L_j \mathscr{P}_j\|$, and

$$b_8 = 2\overline{\mathrm{Lip}}(\mathrm{INV}(\Upsilon_{[m]}))\|A_{0 \sim n-1}\|\|B_n^{\mathsf{T}} P\|.$$

If the parameters satisfy the following equation

$$\ell > \max \left\{1, \frac{2mb_7}{\kappa_2}, \frac{(b_6 + b_8)^2}{2\kappa_1 \kappa_2}\right\}, \tag{6.51}$$

then $\frac{\mathrm{d}V}{\mathrm{d}t} < 0, \forall t > 0$, i.e., the observational error and the tracking error are asymptotically stable. Corollary 6.1 is thus proved.

6.4 Self-Healing Fault-Tolerant Control With Sensor Faults

6.4.1 Definitions and Conditions for Fault Tolerance

When the FAS model (6.1) encounters sensor faults, the aforementioned bounded stability conclusion no longer holds. Therefore, a safer SHFTC framework should be developed. Proposition 6.2 is a sufficient condition for the SHFTC design of the original FAS (6.1).

Definition 6.2 (SHFTC [3, 113]) The SHFTC handles faults in dynamic systems through self-perception, self-diagnosis, self-decision-making, and self-repairing methods. It generally consists of two parts: fault diagnosis and controller reconfiguration, which fall under the category of AFTC. The traditional AFTC focuses on fault estimation and fault compensation, while the SHFTC emphasizes fault isolation and fault elimination. The SHFTC technology proposed in this chapter first self-diagnoses the sensor faults of the original system, then eliminates the faults, and finally restores the sensor fault signals by reconfiguring the decoupled system state.

Assumption 6.2 The FAS (6.1) has at most d sensor faults when $t > t_0$. The remaining fixed set of ideal sensors is U, and the cardinality of U satisfies $|U| \geq m - d$.

Remark 6.2 Assumption 6.2 restricts the number of sensor faults. Many classical sensor fault FTC methods [65, 107] require that the faults be bounded or differentiable, but the sensor faults in this chapter can be unbounded or non-differentiable. Moreover, $t = t_2(t_2 > t_0)$ is the stable time for trajectory tracking, so sensor faults may occur during the transient process, and the observation error and tracking error at $t = t_0$ may not have fully converged. Sensor faults during the transient process are fatal to many robust FTC methods.

Definition 6.3 (Ideal Sensor Fault Tolerance) If within a finite-time interval $[t_0, t_1]$, the control input and the measurement output disrupted by sensor faults can uniquely determine the system state at $t = t_0$, then the sensor faults are regarded to be ideally fault-tolerant.

Remark 6.3 Unlike the existing sensor FTC approaches [65, 129], Definition 6.3 focuses on determining the ideal state observations from the perspective of redundant observability. At this time, the observation performance under sensor faults needs to be consistent with that of the fault-free system.

According to Theorem 6.1, the comprehensive observation error of the fault model (6.8) can be divided into the following two parts:

$$\Upsilon_{[m]}(x^{(0 \sim n-1)}) - \hat{\xi} = \mathfrak{e} + \mathfrak{r}_{[m]}, \tag{6.52}$$

where $\mathfrak{e}$ represents the observation error without sensor faults, and $\mathfrak{r}_{[m]}$ is the residual caused by sensor faults. The ideal sensor set U corresponds to $\|\mathfrak{r}_U\| = 0$, and the observation error without sensor faults $\mathfrak{e}$ satisfies

$$\|\mathfrak{e}\| \leq \gamma(t) \triangleq \max \left\{ \max_{j \in [m]} \{m\beta_j \exp(-\alpha_j \ell t)\}, \frac{mM_1}{\ell} \right\}. \tag{6.53}$$

Theorem 6.2 *When the FAS model (6.1) encounters d sensor faults and assuming $\mathfrak{e} = 0$, the sensor faults are ideally fault-tolerant if and only if the FAS model (6.1) is 2d-redundant observable.*

Proof Since ξ spans the entire observable space of the FAS model (6.1), the FTC problem can be equivalently transformed into a signal recovery problem, i.e.,

$$\Upsilon_{[m]}(x^{(0 \sim n-1)}) = \hat{\xi} + \mathfrak{r}_{[m]}, \mathfrak{r} = [\mathfrak{r}_1^{\mathsf{T}}, \mathfrak{r}_2^{\mathsf{T}}, \cdots, \mathfrak{r}_m^{\mathsf{T}}]^{\mathsf{T}}, \tag{6.54}$$

where

$$\mathfrak{r} \in \mathscr{S}_d \triangleq \{\mathfrak{r} \mid \mathfrak{r} \in \mathbb{R}^{\sum_{j \in [m]} r_j} : \mathrm{SUM}\,(\mathfrak{r}_j \neq \mathbf{0}_{r_j}, j \in [m]) \leq d\}, \tag{6.55}$$

and SUM $(\cdot)$ indicates the number of elements that meet the inner conditions. The signal recovery problem is manifested as whether the sequence $x^{(0 \sim n-1)}$ can be recovered from $\hat{\xi}$ which has an error of d sparse vectors.

(1) Prove the sufficiency by contradiction. If the sensor faults are not ideally fault-tolerant, then there exist $x_{a1}^{(0 \sim n-1)} \neq x_{a2}^{(0 \sim n-1)}$ and $\mathfrak{r}_{a1}, \mathfrak{r}_{a2} \in \mathscr{S}_d$ such that

$$\Upsilon_{[m]}(x_{a1}^{(0 \sim n-1)}) - \mathfrak{r}_{a1} = \Upsilon_{[m]}(z_{a2}^{(0 \sim n-1)}) - \mathfrak{r}_{a2}. \tag{6.56}$$

It follows that

$$\Upsilon_{[m]}(x_{a1}^{(0 \sim n-1)}) - \Upsilon_{[m]}(x_{a2}^{(0 \sim n-1)}) = \mathfrak{r}_a, \tag{6.57}$$

where $\mathfrak{r}_a = \mathfrak{r}_{a1} - \mathfrak{r}_{a2} \in \mathscr{S}_{2d}$. Therefore, there exists a sensor set J, $|J| = m - 2d$ such that

$$\Upsilon_J(x_{a1}^{(0 \sim n-1)}) - \Upsilon_J(x_{a2}^{(0 \sim n-1)}) = \mathbf{0}_{\sum_{j \in J} r_j}, \tag{6.58}$$

where $\mathbf{0}_{\sum_{j \in J} r_j}$ is a zero vector of dimension $\sum_{j \in J} r_j$. This contradicts the fact that Υ_J is an injective immersion mapping, meaning that the FAS model (4.1) is incompatible with the $2d$-redundant observable property.

(2) Prove the necessity by contradiction. If the mapping Υ_J is not an injective immersion mapping, and the cardinality of J is $|J| = m - 2d$, then there exists $x_a^{(0 \sim n-1)} \neq \mathbf{0}_n$ such that

$$\Upsilon_J(x_a^{(0 \sim n-1)}) = \mathbf{0}_{\sum_{j \in J} r_j}. \tag{6.59}$$

This means that there exists $\mathfrak{r}_a \in \mathscr{S}_{2d}$ such that

$$\Upsilon_{[m]}(x_a^{(0 \sim n-1)}) = \mathfrak{r}_a. \tag{6.60}$$

Let $\mathfrak{r}_a = \mathfrak{r}_{a1} - \mathfrak{r}_{a2}, \mathfrak{r}_{a1}, \mathfrak{r}_{a2} \in \mathscr{S}_d$, it follows that

$$\Upsilon_{[m]}(x_a^{(0 \sim n-1)}) - \mathfrak{r}_{a1} = \Upsilon_{[m]}(\mathbf{0}_n) - \mathfrak{r}_{a2}. \tag{6.61}$$

This is in contradiction to the fact that sensor faults are ideally fault-tolerant. Theorem 6.2 is thus proved.

Remark 6.4 Theorem 6.2 analyzes the relationship between redundant observability and sensor fault tolerance in FASs. Moreover, the research results also reveal the maximum number of sensor faults that a nonlinear FAS can tolerate, which has guiding significance for the sensor layout of practical systems.

6.4.2 Fault Diagnosis Methods

The fault detection residual under the full mapping is

$$\mathfrak{R}_{[m]} = \Upsilon_{[m]}\left(\text{INV}(\Upsilon_{[m]})(\hat{\xi})\right) - \hat{\xi}. \tag{6.62}$$

This residual signal can only determine whether there are sensor faults in the FAS, but it cannot realize the fault isolation. Therefore, a residual based on the subset sensor set I is designed as

$$\mathfrak{R}_I = \Upsilon_I\left(\text{INV}(\Upsilon_I)(\hat{\xi}_I)\right) - \hat{\xi}_I, \tag{6.63}$$

where $I \subset [m]$, and its cardinality is $|I| = m - d$.

Theorem 6.3 *For the 2d-redundant observable FAS (6.1) that meets Assumption 6.2, if (6.64) holds, then there must exist at least one fault in the sensor set I*

$$\|\mathfrak{R}_I\| > \overline{\text{Lip}}\left(\Upsilon_I \circ \text{INV}(\Upsilon_I) - \mathscr{I}\right)\gamma(t), \tag{6.64}$$

where $\mathscr{I}$ is the identity mapping on the space $\mathbb{R}^{\sum_{j\in I} r_j}$, and $\circ$ represents the operation of mapping multiplication. The ideal sensor set U satisfies

$$U \subset \bigcup\left\{I \mid I \subset [m] : \|\mathfrak{R}_I\| \leq \overline{\text{Lip}}\left(\Upsilon_I \circ \text{INV}(\Upsilon_I) - \mathscr{I}\right)\gamma(t)\right\}. \tag{6.65}$$

Conversely, if (6.64) does not hold, then there exist positive constants t_3 and M_4 such that

$$\|x^{(0\sim n-1)} - \hat{X}_I\| \leq M_4/\ell, \ \forall t > t_3, \tag{6.66}$$

where $\hat{X}_I$ is the system observation determined by the sensor set I.

Proof According to (6.63), analyzing the residual signals of the selected index gives

$$\|\mathfrak{R}_I\| = \|\Upsilon_I\left(\mathrm{INV}(\Upsilon_I)(\hat{\xi}_I)\right) - \hat{\xi}_I\|$$

$$= \|\Upsilon_I\left(\mathrm{INV}(\Upsilon_I)(\hat{\xi}_I)\right) - \hat{\xi}_I - [\Upsilon_I\left(\mathrm{INV}(\Upsilon_I)(\Upsilon_I(x^{(0\sim n-1)}))\right)$$

$$- \Upsilon_I(x^{(0\sim n-1)})]\|$$

$$= \|(\Upsilon_I \circ \mathrm{INV}(\Upsilon_I) - \mathscr{I})(\hat{\xi}_I) - (\Upsilon_I \circ \mathrm{INV}(\Upsilon_I) - \mathscr{I})(\Upsilon_I(x^{(0\sim n-1)}))\|$$

$$\leq \overline{\mathrm{Lip}}\left(\Upsilon_I \circ \mathrm{INV}(\Upsilon_I) - \mathscr{I}\right)\|\hat{\xi}_I - \Upsilon_I(x^{(0\sim n-1)})\|$$

$$\leq \overline{\mathrm{Lip}}\left(\Upsilon_I \circ \mathrm{INV}(\Upsilon_I) - \mathscr{I}\right)\|e_I + \mathfrak{r}_I\|. \tag{6.67}$$

If there are no faults within the sensor set I, then $\mathfrak{r}_I = 0$, i.e.,

$$\|\mathfrak{R}_I\| \leq \overline{\mathrm{Lip}}\left(\Upsilon_I \circ \mathrm{INV}(\Upsilon_I) - \mathscr{I}\right)\gamma(t). \tag{6.68}$$

This verifies the effectiveness of the detection strategy (6.64). If the residual signal does not trigger an alarm, then (6.68) holds. This does not mean that the FAS is necessarily fault-free, because the fault could be very small. In fact,

$$\mathfrak{r}_I \in \mathscr{S}_d^I \triangleq \{\mathfrak{r}_I|\mathfrak{r}_I \in \mathbb{R}^{\mathcal{N}_I} : \mathrm{SUM}\,(\mathfrak{r}_j \neq \mathbf{0}_{r_j}, j \in I) \leq d\}. \tag{6.69}$$

The d-redundant observability means that for any sensor set $J \subset I$ with $|J| = m - 2d$, it holds $\mathfrak{r}_J = 0$. According to (6.52), the following can be calculated as

$$\|\Upsilon_I\left(\mathrm{INV}(\Upsilon_I)(\hat{\xi}_I)\right) - \hat{\xi}_I\| \geq \|\Upsilon_J\left(\mathrm{INV}(\Upsilon_I)(\hat{\xi}_I)\right) - \hat{\xi}_J\|$$

$$= \|\Upsilon_J\left(\mathrm{INV}(\Upsilon_I)(\hat{\xi}_I)\right) - \Upsilon_J(x^{(0\sim n-1)}) - e_J\|$$

$$\geq \underline{\mathrm{Lip}}(\Upsilon_J)\|\mathrm{INV}(\Upsilon_I)(\hat{\xi}_I) - x^{(0\sim n-1)}\| - \|e_J\|$$

$$\geq \underline{\mathrm{Lip}}(\Upsilon_J)\|x^{(0\sim n-1)} - \hat{X}_I\| - \gamma(t). \tag{6.70}$$

By combining (6.68) and (6.70), it can be obtained that

$$\|x^{(0\sim n-1)} - \hat{X}_I\| \leq \frac{\overline{\mathrm{Lip}}\left(\Upsilon_I \circ \mathrm{INV}(\Upsilon_I) - \mathscr{I}\right) + 1}{\min_{J \subset I} \underline{\mathrm{Lip}}(\Upsilon_J)}\gamma(t). \tag{6.71}$$

Obviously, (6.71) demonstrates the significance of $2d$-redundant observability, as $2d$-redundant observability ensures the bi-Lipschitz property of $\Upsilon_J, J \subset I$. Therefore, (6.66) can easily hold. Theorem 6.3 has been proved.

Remark 6.5 If the FAS (6.1) is only d-redundant observable, then the residual signal (6.62) can only guarantee the fault detection function. Based on Theorem 6.2, a d-redundant observable FAS may not be capable of tolerating d sensor faults.

6.4.3 Sensor-Based Control Strategies

Based on Theorem 6.3, a continuously updated sensor set $\iota(t)$ with $|\iota(t)| = m - d$ is desgined as follows. If

$$\|\mathfrak{R}_{\iota(t)}\| > \overline{\text{Lip}} \left(\Upsilon_{\iota(t)} \circ \text{INV}(\Upsilon_{\iota(t)}) - \mathscr{I} \right) \gamma(t), \tag{6.72}$$

then the sensor set is updated to $\iota(t^+)$, where t^+ represents the next calculation time after t. The update of the sensor set will iterate through all combinations of the index set $I \subset [m], |I| = m - d$. Finally, the SHFTC framework based on the redundancy principle is designed as

$$
\textbf{Full observation} : \dot{\hat{\xi}}_j =
\begin{bmatrix}
\hat{\xi}_{j,2} \\
\vdots \\
\hat{\xi}_{j,r_j} \\
F_o^j(\hat{\xi}_j)
\end{bmatrix}
$$

$$
+
\begin{bmatrix}
\ell_j L_{j,1} \\
\ell_j^2 L_{j,2} \\
\vdots \\
\ell_j^{r_j} L_{j,r_j}
\end{bmatrix}
(y_j - \hat{\xi}_{j,1}) +
\begin{bmatrix}
0 \\
\vdots \\
0 \\
G_o^j(\hat{\xi}_j)
\end{bmatrix}
u', j \in [m]
$$

Fault diagnosis : If (6.72) holds, $\iota(t)$ updates

Ideal observation : $\hat{X}_{\iota(t)} = \text{INV}(\Upsilon_{\iota(t)})(\hat{\xi}_{\iota(t)})$

Controller : $u' = -G^{-1}(\hat{X}_{\iota(t)}) \left[F(\hat{X}_{\iota(t)}) - x_r^{(n)} + A_{0\sim n-1} \left(\hat{X}_{\iota(t)} - x_r^{(0\sim n-1)} \right) \right]$

$$\tag{6.73}$$

where $\hat{\xi}_{\iota(t)}, \hat{X}_{\iota(t)}$ are the observations under the sensor set $\iota(t)$. The update of the sensor set can be designed in a natural order or in a descending order based on the relative order.

Theorem 6.4 *For the 2d-redundant observable FAS model (6.1) under the SHFTC framework (6.73), if Assumption 6.2 holds, then there exist positive constants t_4, t_5, M_5, and M_6 such that the observation error and the tracking error are ultimately uniformly bounded, i.e.,*

$$
\begin{cases}
\|x^{(0\sim n-1)} - \hat{X}_{\iota(t)}\| \leq M_5/\ell, \; \forall t > t_4, \\
\|x - x_r\| \leq M_6/\ell, \; \forall t > t_5,
\end{cases}
\tag{6.74}
$$

Furthermore, the tracking error converges to a neighborhood near the origin. For parameters $\ell \gg 1$, the parameter ℓ can result in a tiny tracking error.

Proof Compared with Theorem 6.1, Theorem 6.4 includes a dynamic sensor set update law $\iota(t)$, which is a dynamic set that only occurs during the fault diagnosis process. For a $2d$-redundant observable FAS, it is easy to determine an available sensor set where there may exist incipient faults. These incipient faults have an impact smaller than that of noise, so they can be reasonably ignored. At this time, the available set can be regarded as the ideal set U. When a sensor fails, using Theorem 6.3 and (6.72), the ideal sensor set can be quickly determined. The constraint of compact sets can avoid the system divergence phenomenon during the fault diagnosis process. After fault isolation, the ideal set U is fixed as the sensor set used for state observation. Then, the following Lyapunov function is selected to prove stability

$$V = \sum_{j \in U} \varepsilon_j^\mathsf{T} \mathscr{P}_j \varepsilon_j + E^\mathsf{T} P E. \tag{6.75}$$

The remaining proof is similar to Theorem 6.1.

Corollary 6.2 *For the 2d-redundant observable FAS (6.38) with sensor faults, the proposed SHFTC framework (6.73) can ensure that the ultimate error dynamics is asymptotically stable.*

Proof The proof process is similar to Theorem 6.4 and Corollary 6.1. Many practical objects can be modeled as (6.38), such as inverted pendulum models, robotic arm models, etc.

Remark 6.6 The proposed SHFTC framework can provide a systematic analysis of redundancy observability and decouple the impact of sensor faults on system observation and control. Compared with the existing sensor FTC methods [38, 79, 106, 116, 129], it can handle more general faults. More importantly, the proposed SHFTC framework does not rely on the error system in the steady-state process, and the fault diagnosis strategy (6.72) can detect sensor faults during the transient process. Therefore, this framework is more versatile and can also be applied in multi-model switching environments.

6.5 Experimental Validation

To illustrate the effectiveness of the proposed SHFTC framework, the following third-order FAS model is considered as

$$\begin{cases} x^{(3)} = 3 - 3\ddot{x} - 2\dot{x} - \sin(\ddot{x} + 2\dot{x} - 2) + u, \\ y_1 = x + s_1 + v_1, \\ y_2 = x + s_2 + v_2, \\ y_3 = \dot{x} + s_3 + v_3, \\ y_4 = 2\dot{x} + 4x - 2 + s_4 + v_4. \end{cases} \tag{6.76}$$

Table 6.1 Four mappings Υ_ι^{-1}, $|\iota| = 3$

Mapping	Specific form
$\Upsilon_{\{2,3,4\}}^{-1} : \mathbb{R}^7 \to \mathbb{R}^3$	$[\xi_{2,1}, \xi_{3,1}, 0.5\xi_{4,2} - 2\xi_{3,1}]^\mathsf{T}$
$\Upsilon_{\{1,3,4\}}^{-1} : \mathbb{R}^7 \to \mathbb{R}^3$	$[0.25\xi_{4,1} - 0.5\xi_{3,1} + 0.5, \xi_{1,2}, 0.5\xi_{4,2} - 2\xi_{1,2}]^\mathsf{T}$
$\Upsilon_{\{1,2,4\}}^{-1} : \mathbb{R}^8 \to \mathbb{R}^3$	$[\xi_{1,1}, \xi_{2,2}, 0.5\xi_{4,2} - 2\xi_{1,2}]^\mathsf{T}$
$\Upsilon_{\{1,2,3\}}^{-1} : \mathbb{R}^8 \to \mathbb{R}^3$	$[\xi_{1,1}, \xi_{2,2}, \xi_{3,2}]^\mathsf{T}$

where s_j, $j \in [4]$ represents possible sensor faults, and the noise v_j, $j \in [4]$ follows a uniform distribution within the range of $[-0.005, 0.005]$. The control task is to make x track the reference signal $x_r = \sin t$. The observable space corresponding to each measurement is

$$
\begin{cases}
\xi_1 = \Upsilon_1(x^{(0\sim 2)}) = [x, \dot{x}, \ddot{x}]^\mathsf{T}, \\
\xi_2 = \Upsilon_2(x^{(0\sim 2)}) = [x, \dot{x}, \ddot{x}]^\mathsf{T}, \\
\xi_3 = \Upsilon_3(x^{(0\sim 2)}) = [\dot{x}, \ddot{x}]^\mathsf{T}, \\
\xi_4 = \Upsilon_4(x^{(0\sim 2)}) = [2\dot{x} + 4x - 2, 2\ddot{x} + 4\dot{x}]^\mathsf{T}.
\end{cases}
\tag{6.77}
$$

The relative orders are 3, 3, 2, and 2, and the FAS model (6.76) is 2-redundant observable. According to Theorem 6.2, this system can tolerate only one sensor fault. Clearly, if both sensor sets $\{1, 2\}$ are faulty, the control task cannot be completed. In the experiment, the mappings Υ_ι^{-1} is shown in Table 6.1.

The simulated injected fault is $s_2 = t \sin t$, where $t > t_f$, and t_f is the time of fault occurrence. The parameter selection is $A_{0\sim 2} = [2, 4, 3]$, $\ell = 20$, $L_1 = L_2 = [2, 4, 3]$, and $L_3 = L_4 = [5, 2]$. The system state is within the compact set, i.e., $x^{(0\sim 2)} \in \Omega_x \triangleq \{\varsigma \mid \varsigma \in \mathbb{R}^3 : \|\varsigma\|_\infty \leq 10\}$, $\xi_j \in \Omega_j \triangleq \{\varsigma\varsigma \in \mathbb{R}^{10} : \|\varsigma\|_\infty \leq 20\}$.

6.5.1 An Experiment in a Steady-State Process

The fault occurs at $t_f = 20$s, and the ideal fault detection strategy is

$$
\|\mathfrak{R}_{\iota(t)}\| > \overline{\mathrm{Lip}} \left(\Upsilon_{\iota(t)} \circ \mathrm{INV}(\Upsilon_{\iota(t)}) - \mathscr{I} \right) \gamma(t).
\tag{6.78}
$$

However, the specific numerical values of the ideal detection logic are difficult to determine. Through fault-free Monte Carlo simulation, it can be concluded that $\sup_{t\geq 0} \|\gamma(t)\| < 0.1, t > t_2$. After combining the information of the compact sets Ω_x and Ω_J, a constant greater than the ideal threshold is selected as the experimental threshold. Therefore, the simplified fault detection strategy is $\|\mathfrak{R}_{\iota(t)}\| > 2.1$. The experimental results are shown in Figs. 6.2, 6.3, and 6.4. The FAS is unstable without FTC, and the proposed SHFTC framework can quickly detect and isolate faults. Since sensor faults can cause significant damage to the observation and control, the residual signal is very sensitive to sensor faults.

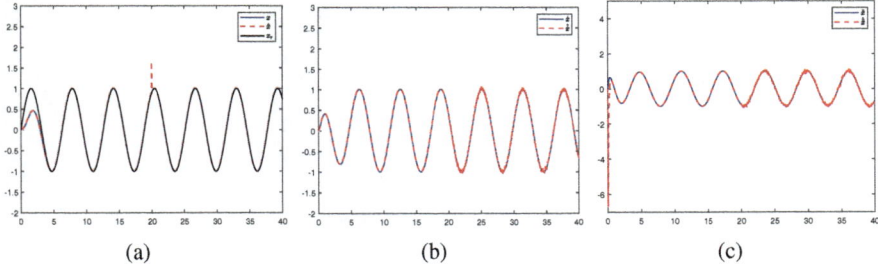

Fig. 6.2 The signal trajectories under the SHFTC in a steady-state process

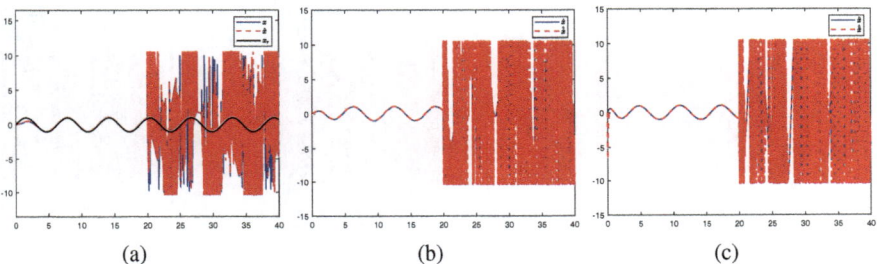

Fig. 6.3 The signal trajectories without the FTC in a steady-state process

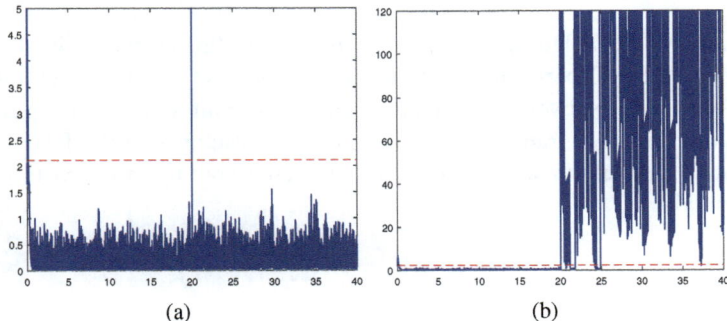

Fig. 6.4 The residual signals under the two strategies in a steady-state process: (**a**) SFHFTC; (**b**) without FTC

6.5.2 An Experiment in a Transient Process

The fault occurs at $t_f = 0.1$s. Similarly, the simplified fault detection strategy is $\|\mathfrak{R}_{i(t)}\| > 7$. The experimental results are shown in Figs. 6.5 and 6.6. The SHFTC framework proposed in this chapter can detect and isolate faults during the transient process. The fault-tolerant performance of the transient process is inferior to that of the steady-state process because this SHFTC framework utilizes a simplified detection threshold and does not utilize the actual time-varying signal $\gamma(t)$. The signal in the transient process changes rapidly, and the simplified detection logic

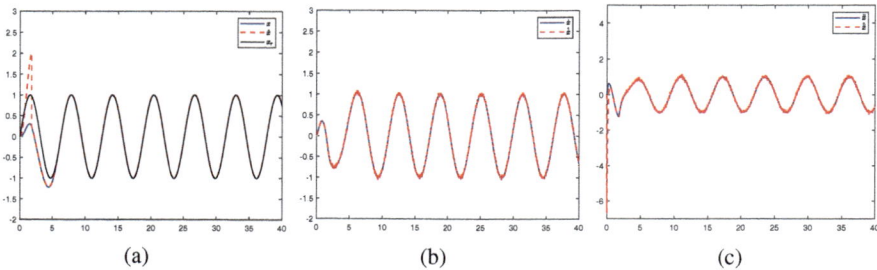

Fig. 6.5 The signal trajectories under the SHFTC in a transient process

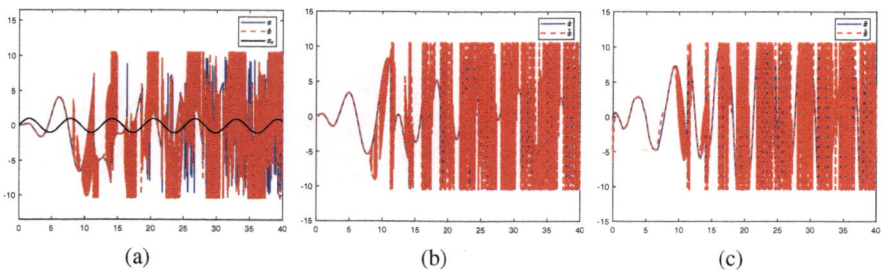

Fig. 6.6 The signal trajectories without the FTC in a transient process

will slow down the fault diagnosis speed. If an accurate time-varying signal $\gamma(t)$ can be obtained, the FTC performance in the transient process will be further improved. Finally, due to the relatively weak requirement for prior information of sensor faults, the proposed SHFTC framework has significant advantages in the FTC fields of novel faults and serious faults, which are not achievable by most existing sensor FTC methods.

6.6   Notes and References

This chapter proposes an SHFTC framework based on system redundancy feature extraction for FAS models, to address sensor faults during steady-state or transient processes. The relationship between redundant measurement and redundant observability is established, and a necessary and sufficient condition detail the relationship between redundant observability and sensor fault tolerance. The redundant observable structure helps determine the maximum number of sensor faults that a system can tolerate. Moreover, the SHFTC framework in this chapter has a relatively weak conservatism and has broad application prospects in the FTC field of sudden and catastrophic sensor faults.

Part II
FTC for FASs Against Multiple Faults

Chapter 7
Integrated FTC for FASs Against Actuator and Sensor Faults

This chapter presents three active fault-tolerant control (AFTC) frameworks for fully actuated systems (FASs) with actuator faults, sensor faults and measurement noise. Through the redundant observability, actuator and sensor faults are tolerated. The AFTC general framework based on traditional fusion observers can merely give a bounded error system. To further suppress noise, a saturation fusion strategy and a dead-zone fusion strategy are applied into the AFTC redesign, respectively. Especially in a linear FAS structure, the noise suppression performance of the two redesigned AFTC frameworks are proved to be superior. At last, some comparative cases experimentally illustrate fault tolerance and noise suppression performance.

7.1 Overview and Scope of Integrated Fault-Tolerant Control Systems

Actuator faults will affect the control performance, while sensor faults will disrupt the state measurements. When both types of faults exist simultaneously, it poses higher requirements for the design of the fault-tolerant control (FTC) for nonlinear systems. Chapters 2 to 6 all study the FTC problem of nonlinear FASs with a single fault type. Due to the high requirements of the FAS theory for model accuracy and idealness, the design of FTC that handles both types of faults simultaneously is indispensable. Moreover, the measurement equations of nonlinear systems are often noisy, and noise can seriously affect the observation and control of high-order systems. The low-power fault-tolerant control framework in Chap. 4 only suppresses a special type of high-frequency noise, but does not achieve good suppression effects for general noise. Under the deterministic FAS structure, even if the observer and controller that completely eliminate noise are physically impossible to achieve, appropriate noise suppression strategies still need to be studied.

© The Author(s) 2026

D. Zhou. M. Cai, *Fault-Tolerant Control for Fully Actuated Systems*,
https://doi.org/10.1007/978-981-95-0691-0_7

In the case of FASs with both actuator and sensor faults, this chapter proposes three AFTC frameworks. Based on the controllability and observability of FASs, these integrated FTC designs are carried out from two aspects: actuator fault compensation and sensor fault redundancy. Starting from each measurable subspace, the first AFTC structure provides a fusion estimation of high-order states and actuator faults. Under the fault detection logic based on the dynamic sensor set, this AFTC design can effectively handle sensor faults with unbounded amplitudes. Considering the influence of noise on high-order systems, the next two AFTC structures respectively utilize saturation design and dead-zone design strategies to suppress noise, and the noise suppression performance of these two novel AFTC structures in linear FASs has been mathematically proven to be better.

7.2 Problem Formulation and Key Insights

Consider a class of FASs with both actuator and the sensor faults:

$$\begin{cases} x^{(n)} = F(x^{(0\sim n-1)}) + G(x^{(0\sim n-1)})u + \delta, \\ y_j = H_j(x^{(0\sim n-1)}) + s_j + v_j, \ j \in [m], \end{cases} \tag{7.1}$$

where $x, u, y_j \in \mathbb{R}$ are the systems state, the control input, and the measurement output. The high-order state is concentrated in a compact set, i.e., $x^{(0\sim n-1)} \in \Omega_x \triangleq \{\varsigma \mid \varsigma \in \mathbb{R}^n : \|\varsigma\|_\infty \le \Xi_x\}$, where Ξ_x is a positive constant. The nonlinear functions $F(\cdot), G(\cdot), H_j(\cdot), j \in [m]$ are smooth, and the controlled object (6.1) is fully actuated, i.e., $\det G(x^{(0\sim n-1)}) \ne 0$ or ∞, for all $x^{(0\sim n-1)} \in \Omega_x$. The measurement noise $v = [v_1, v_2, \cdots, v_m]^\mathsf{T}$ has a certain upper bound $\bar{v}$, and $\delta(t)$ is a possible actuator fault, in the form of [138]

$$\delta = -\rho G(x^{(0\sim n-1)})u + u_f, \tag{7.2}$$

where ρ and u_f represent the time-varying gain attenuation fault and deviation fault, respectively. $s_j, j \in [m]$ represent the sensor faults. The control task of this chapter is to design a fault-tolerant tracking controller when both faults and noise exist, so that x can track the smooth reference signal x_r.

Assumption 7.1 The dynamics and faults of the FAS model (7.1) should satisfy the following conditions:

1. The functions $F(\cdot), G(\cdot)$ are Lipschitz, and their partial derivatives are bounded.
2. The reference signal x_r satisfies $\sup_{t \ge 0} \|x_r^{(0\sim n)}\| < \infty$.
3. δ satisfies $\sup_{t \ge 0} \|[\delta(t), \dot{\delta}(t)]\| < \infty$.
4. The system can encounter at most d sensor faults, which may be of unbounded amplitude or non-differentiable.

Lemma 7.1 ([28]) *The nonlinear FAS (7.1) is completely controllable. Particularly, when the nonlinear system (7.1) has no faults or noise, and the high-order states $x^{(0\sim n-1)}$ can be directly measured, a parameterized controller*

$$u = -G^{-1}(x^{(0\sim n-1)}) \left[A_{0\sim n-1} \left(x^{(0\sim n-1)} - x_r^{(0\sim n-1)} \right) + F(x^{(0\sim n-1)}) + x_r^{(n)} \right] \tag{7.3}$$

can give a globally asymptotically stable system

$$\dot{E} = \Phi_A E, \tag{7.4}$$

where $E = [e_1, e_2, \cdots, e_n]^\mathsf{T} \triangleq e^{(0\sim n-1)} = x^{(0\sim n-1)} - x_r^{(0\sim n-1)}$,

$$\Phi_A = \begin{bmatrix} 0 & 1 & 0 & \cdots & 0 \\ 0 & 0 & 1 & \cdots & 0 \\ \vdots & & & \ddots & \vdots \\ 0 & 0 & 0 & \cdots & 1 \\ -A_0 & -A_1 & -A_2 & \cdots & -A_{n-1} \end{bmatrix} \tag{7.5}$$

is a Hurwitz matrix.

However, the presence of faults and noise makes it difficult to easily obtain the controller (7.3). As long as the actuator fault δ does not have a significant sudden change, many classical fault estimators can compensate for the actuator fault, but traditional observers based on augmented systems cannot estimate sensor faults s_j, $j \in [m]$ without amplitude constraints.

Since the high-order state information is determined by m nonlinear measurement equations, it is necessary to decompose the observable space of the FAS model. Inspired by Chap. 6, the redundancy observability and m observable structures of the FAS model should be first analyzed. The following two vector fields are defined as

$$\begin{cases} \mathscr{F} = [\dot{x}, \ddot{x}, \cdots, x^{(n-1)}, F(x^{(0\sim n-1)})]^\mathsf{T}, \\ \mathscr{G} = [0, 0, \cdots, 0, G(x^{(0\sim n-1)})]^\mathsf{T}, \end{cases} \tag{7.6}$$

then there exist m positive integers r_j, where $j \in [m]$, such that for all $t > 0$ and $k \in [r_j - 2]$, it holds that

$$\mathscr{L}_\mathscr{G} \mathscr{L}_\mathscr{F}^{r_j-1} H_j(x^{(0\sim n-1)}) \neq 0, \ \mathscr{L}_\mathscr{G} \mathscr{L}_\mathscr{F}^{k} H_j(x^{(0\sim n-1)}) = 0. \tag{7.7}$$

Define

$$
\xi_j = \begin{bmatrix} \xi_{j,1} \\ \xi_{j,2} \\ \vdots \\ \xi_{j,r_j} \end{bmatrix} = \begin{bmatrix} \mathcal{L}_{\mathscr{F}} H_j(x^{(0\sim n-1)}) \\ \mathcal{L}_{\mathscr{F}}^2 H_j(x^{(0\sim n-1)}) \\ \vdots \\ \mathcal{L}_{\mathscr{F}}^{r_j-1} H_j(x^{(0\sim n-1)}) \end{bmatrix} \triangleq \Upsilon_j(x^{(0\sim n-1)}), \tag{7.8}
$$

and there exists a differential homeomorphism $\hbar_j : \mathbb{R}^n \to \mathbb{R}^{r_j} \times \mathbb{R}^{n-r_j}$, i.e., $[\xi_j^{\mathsf{T}}, \eta_j^{\mathsf{T}}]^{\mathsf{T}} = \hbar_j(x^{(0\sim n-1)})$, such that the controlled system (7.1) under the jth measurement can be rewritten as [39, 50]

$$
\dot{\xi}_j = \begin{bmatrix} \dot{\xi}_{j,1} \\ \dot{\xi}_{j,2} \\ \vdots \\ \dot{\xi}_{j,r_j} \end{bmatrix} = \begin{bmatrix} \xi_{j,2} \\ \vdots \\ \xi_{j,r_j} \\ F_o^j(\xi_j) + \delta \end{bmatrix} + \begin{bmatrix} 0 \\ \vdots \\ 0 \\ G_o^j(\xi_j) \end{bmatrix} u, \tag{7.9a}
$$

$$
\dot{\eta}_j = F_{uo}^j(\xi_j, \eta_j) + G_{uo}^j(\xi_j, \eta_j)u + K_{uo}^j(\xi_j, \eta_j)\delta, \tag{7.9b}
$$

$$
y_i = \xi_{j,1} + s_j + v_j, \tag{7.9c}
$$

where $F_o^j(\cdot), G_o^j(\cdot), F_{uo}^j(\cdot), G_{uo}^j(\cdot), K_{uo}^j(\cdot)$ are Lipschitz functions, $\xi_j, j \in [m]$ are in compact sets, i.e., $\xi_j \in \Omega_j \triangleq \{\varsigma \mid \varsigma \in \mathbb{R}^{r_j} : \|\varsigma\|_\infty \leq \varXi_j\}$, and $\varXi_j, j \in [m]$ are positive constants. (7.9a) represents the observable structure corresponding to the variable y_j, and $\xi = [\xi_1^{\mathsf{T}}, \xi_2^{\mathsf{T}}, \cdots, \xi_m^{\mathsf{T}}]^{\mathsf{T}} \triangleq \Upsilon_{[m]}(x^{(0\sim n-1)}) \in \mathbb{R}^{\sum_{j \in [m]} r_j}$ is the overall observable structure of the system (7.1).

Lemma 7.2 (Observability and Redundant Observability of FASs [55]) *If the mapping $\Upsilon_{[m]}$ is an injective immersion mapping, then the nonlinear FAS model (7.1) is completely observable. Define a mapping $\Upsilon_I, \forall I \subset [m], |I| = m - d,$*

$$
\xi_I = [\xi_{I_1}^{\mathsf{T}}, \xi_{I_2}^{\mathsf{T}}, \cdots, \xi_{I_{m-d}}^{\mathsf{T}}] \triangleq \Upsilon_I(x^{(0\sim n-1)}) \in \mathbb{R}^{\sum_{j \in I} r_j}, \tag{7.10}
$$

if for $\forall I \subset [m]$, the mapping Υ_I is an injective immersion mapping, then the nonlinear FAS model (7.1) is d-redundant observable.

The general observability, namely 0-redundant observability, is the prerequisite for the design of ideal system observers and controllers. When the system is disrupted by amplitude-unbounded sensor faults, since the sensor faults may not be compensable, the redundant observability of the system will become the core of FTC design.

7.3   General Fault Tolerance Design Strategies

Considering the limited performance of passive fault-tolerant control (PFTC), in the FAS model (7.1) where both actuators and sensors are faulty simultaneously, a more performant AFTC strategy should be selected. The residual signal is a key step in the design of AFTC, and the residual signal is based on state observation. The actuator fault is estimable, and the residual signal for sensor faults is an important task in the AFTC design of this section. Therefore, this section divides the research into two parts (with or without sensor faults), to facilitate the generation of fault residual and the determination of detection threshold.

7.3.1   Single Fault: Actuator

For each sensor measurement y_j, the corresponding observer is designed as

$$
\begin{cases}
\dot{\hat{\xi}}_j =
\begin{bmatrix}
\hat{\xi}_{j,2} \\
\vdots \\
\hat{\xi}_{j,r_j} \\
F_o^j(\hat{\xi}_j) + \hat{\delta}_j
\end{bmatrix}
+
\begin{bmatrix}
\ell L_{j,1} \\
\ell^2 L_{j,2} \\
\vdots \\
\ell^{r_j} L_{j,r_j}
\end{bmatrix}
(y_j - \hat{\xi}_{j,1}) +
\begin{bmatrix}
0 \\
\vdots \\
0 \\
G_o^j(\hat{\xi}_j)
\end{bmatrix}
u, \\
\dot{\hat{\delta}}_j = \ell^{r_j+1} L_{j,r_j+1}(y_j - \hat{\xi}_{j,1}),
\end{cases}
\tag{7.11}
$$

where $\ell > 1$, $\hat{\xi}_j = [\hat{\xi}_{j,1}, \hat{\xi}_{j,2}, \cdots, \hat{\xi}_{j,r_j}]^\mathsf{T}$ is the estimation of $\xi_j = [\xi_{j,1}, \xi_{j,2}, \cdots, \xi_{j,r_j}]^\mathsf{T}$, $\hat{\delta}_j$ is the sensor fault estimation given by y_j. $L_{j,1}, L_{j,2}, \cdots, L_{j,r_j+1}$ are $r_j + 1$ parameters to be designed, and these parameters can form the following Hurwitz matrix

$$
\Psi_j =
\begin{bmatrix}
-L_{j,1} & 1 \ 0 \cdots 0 \\
-L_{j,2} & 0 \ 1 \cdots 0 \\
\vdots & \ddots \\
-L_{j,r_j} & 0 \ 0 \cdots 1 \\
-L_{j,r_j+1} & 0 \ 0 \cdots 0
\end{bmatrix}
\in \mathbb{R}^{(r_j+1)\times(r_j+1)}.
\tag{7.12}
$$

There are m observers, and the fusion observation is $\hat{\xi} = [\hat{\xi}_1^\mathsf{T}, \hat{\xi}_2^\mathsf{T}, \cdots, \hat{\xi}_m^\mathsf{T}]$. If the controlled system (7.1) is completely observable, i.e., the mapping $\Upsilon_{[m]}$ is invertible, then the observation $\hat{X}$ can be calculated as

$$
\hat{X} = \mathrm{INV}(\Upsilon_{[m]})(\hat{\xi}) \triangleq \mathrm{Sat}_{\Xi_x}\left(\Upsilon_{[m]}^{-1}\left([\mathrm{Sat}_{\Xi_1}\hat{\xi}_1, \mathrm{Sat}_{\Xi_2}\hat{\xi}_2, \cdots, \mathrm{Sat}_{\Xi_m}\hat{\xi}_m]^\mathsf{T}\right)\right),
\tag{7.13}
$$

where $\hat{X} = [\hat{x}_1, \hat{x}_2, \cdots, \hat{x}_n]^\mathsf{T}$ is the estimation of $X = [x_1, x_2, \cdots, x_n]^\mathsf{T} \triangleq x^{(0 \sim n-1)}$, $\Upsilon_{[m]}^{-1}$ is the left inverse mapping of $\Upsilon_{[m]}$ within the compact set, and Sat is the saturation function. The calculation principle of (7.13) is to ensure the validity of the domain through the Lipschitz extension technique [88].

In order to reduce the influence of measurement noise, the fusion fault estimation of δ is designed as

$$\hat{\delta} = \frac{\sum_{j \in [m]} \hat{\delta}_j}{m}. \tag{7.14}$$

Based on the above state observation and fault estimation, an FTC framework for compensating the actuator fault is designed as

$$u_1 = -G^{-1}(\hat{X}) \left[A_{0 \sim n-1} \left(\hat{X} - x_r^{(0 \sim n-1)} \right) + F(\hat{X}) + \hat{\delta} - x_r^{(n)} \right]. \tag{7.15}$$

Lemma 7.3 *For the nonlinear FAS (7.1) with only actuator faults, considering the FTC general design (7.11)–(7.15), the state estimation error, actuator fault estimation error, and tracking error are ultimately uniformly bounded, i.e., there exist positive constants $t_1, t_2, M_1, M_2, M_3, M_4, M_5, M_6$ such that*

$$\begin{cases} \|x^{(0 \sim n-1)} - \hat{X}\| \leq \dfrac{M_1}{\ell^2} + \ell^{n-1} M_2 \bar{v}, \forall t > t_1, \\[2mm] \|\delta - \hat{\delta}\| \leq \dfrac{M_3}{\ell} + \ell^n M_4 \bar{v}, \forall t > t_1, \\[2mm] \|E\| \leq \dfrac{M_5}{\ell} + \ell^n M_6 \bar{v}, \forall t > t_2. \end{cases} \tag{7.16}$$

Furthermore, if there is no measurement noise, i.e., $\bar{v} = 0$, then the aforementioned error can converge to a tiny neighborhood around the origin.

Proof For the measurement y_j, the corresponding observation error is

$$\begin{aligned} \zeta_j &= [\zeta_{j,1}, \zeta_{j,2}, \cdots, \zeta_{j,r_j}, \zeta_{j,r_j+1}]^\mathsf{T} \\ &= [\ell^{r_j}(\xi_{j,1} - \hat{\xi}_{j,1}), \ell^{r_j-1}(\xi_{j,2} - \hat{\xi}_{j,2}), \cdots, \delta - \hat{\delta}_j]^\mathsf{T}, \end{aligned} \tag{7.17}$$

The derivative of ζ_j is

$$\begin{cases} \dot{\zeta}_{j,k} = \ell \zeta_{j,k+1} - \ell L_{j,k} \zeta_{j,1} - \ell^{r_j+1} L_{j,k} v_j, k \in [r_j - 1], \\ \dot{\zeta}_{j,r_j} = \ell \zeta_{j,r_j+1} - \ell L_{j,r_j} \zeta_{j,1} - \ell^{r_j+1} L_{j,r_j} v_j + \varpi_j, \\ \dot{\zeta}_{j,r_j+1} = \dot{\delta} - \ell L_{j,r_j+1} \zeta_{j,1} - \ell^{r_j+1} L_{j,r_j+1} v_j, \end{cases} \tag{7.18}$$

where

$$\varpi_j = \ell[F_o^j(\xi_j) - F_o^j(\hat{\xi}_j) + (G_o^j(\xi_j) - G_o^j(\hat{\xi}_j))u_1]. \tag{7.19}$$

The derivative of high-order tracking error is

$$
\begin{cases}
\dot{e}_i = e_{i+1}, i \in [n-1], \\
\dot{e}_n = \rho + A_{0\sim n-1}(x^{(0\sim n-1)} - \hat{X}) - A_{0\sim n-1}E,
\end{cases}
\tag{7.20}
$$

where

$$
\rho = F(x^{(0\sim n-1)}) - F(\hat{X}) + \delta - \hat{\delta} + [G(x^{(0\sim n-1)}) - G(\hat{X})]u_1.
\tag{7.21}
$$

Therefore, the overall error dynamics is

$$
\begin{cases}
\dot{\zeta}_{j,k} = \ell\zeta_{j,k+1} - \ell L_{j,k}\zeta_{j,1} - \ell^{r_j+1}L_{j,k}v_j, \ j \in [m], k \in [r_j-1], \\
\dot{\zeta}_{j,r_j} = \ell\zeta_{j,r_j+1} - \ell L_{j,r_j}\zeta_{j,1} - \ell^{r_j+1}L_{j,r_j}v_j + \varpi_j, \ j \in [m], \\
\dot{\zeta}_{j,r_j+1} = \dot{\delta} - \ell L_{j,r_j+1}\zeta_{j,1} - \ell^{r_j+1}L_{j,r_j+1}v_j, \ j \in [m], \\
\dot{e}_i = e_{i+1}, i \in [n-1], \\
\dot{e}_n = \rho + A_{0\sim n-1}\tilde{X} - A_{0\sim n-1}e^{(0\sim n-1)},
\end{cases}
\tag{7.22}
$$

where the observation error of the original FAS (7.1) is defined as

$$
\tilde{X} = x^{(0\sim n-1)} - \hat{X} = \text{INV}(\Upsilon_{[m]})(\xi) - \text{INV}(\Upsilon_{[m]})(\hat{\xi}).
\tag{7.23}
$$

Under the condition that the controlled system (7.1) is completely observable, the mapping $\Upsilon_{[m]}$ is an injective immersion mapping, and the mapping $\text{INV}(\Upsilon_{[m]})$ is Lipschitz on the compact set. It follows that

$$
\|\tilde{X}\| \le a_1(\|\xi\| - \|\hat{\xi}\|) \le a_1\|\zeta\|,
\tag{7.24}
$$

where $\zeta = [\zeta_1^\mathsf{T}, \zeta_2^\mathsf{T}, \cdots, \zeta_m^\mathsf{T}]^\mathsf{T}$, and a_1 is a positive constant.

Consider a Lyapunov function

$$
V_1 = \sum_{j=1}^m \zeta_j^\mathsf{T} \mathscr{P}_j \zeta_j + E^\mathsf{T} P E,
\tag{7.25}
$$

where $\mathscr{P}_j$ is the positive definite solution to the following Lyapunov equation

$$
\mathscr{P}_j \Psi_j + \Psi_j^\mathsf{T} \mathscr{P}_j = -\kappa_2 I_{r_j+1},
\tag{7.26}
$$

and P is the positive definite solution to the following Lyapunov equation

$$
P \Phi_A + \Phi_A^\mathsf{T} P = -\kappa_1 I_n
\tag{7.27}
$$

with $\kappa_1, \kappa_2 > 0$. The derivative of V_1 is

$$
\begin{aligned}
\frac{dV_1}{dt} &= \sum_{j=1}^{m} \left\{ \ell \zeta_j^\mathsf{T} (\mathscr{P}_j \Psi_j + \Psi_j^\mathsf{T} \mathscr{P}_j) \zeta_j - 2\ell^{r_j+1} v_j L_j \mathscr{P}_j \zeta_j + 2\dot{\delta} B_{r_j+1}^\mathsf{T} \mathscr{P}_j \zeta_j \right. \\
&\quad \left. + 2\varpi_j \mathfrak{B}_{r_j}^\mathsf{T} \mathscr{P}_j \zeta_j \right\} + E^\mathsf{T} (P\Phi_A + \Phi_A^\mathsf{T} P) E + 2\rho B_n^\mathsf{T} P E + 2A_{0 \sim n-1} \tilde{X} B_n^\mathsf{T} P E \\
&= -\ell \kappa_2 \|\zeta\|^2 - \kappa_1 \|E\|^2 + \sum_{j=1}^{m} \left\{ 2\dot{\delta} B_{r_j+1}^\mathsf{T} \mathscr{P}_j \zeta_j + 2\varpi_j \mathfrak{B}_{r_j}^\mathsf{T} \mathscr{P}_j \zeta_j \right. \\
&\quad \left. - 2\ell^{r_j+1} v_j L_j \mathscr{P}_j \zeta_j \right\} + 2\rho B_n^\mathsf{T} P E + 2A_{0 \sim n-1} \tilde{X} B_n^\mathsf{T} P E,
\end{aligned}
\tag{7.28}
$$

where $L_j = [L_{j,1}, L_{j,2}, \cdots, L_{j,r_j+1}]$, $B_n = [0, 0 \cdots, 0, 1]^\mathsf{T} \in \mathbb{R}^n$, $B_{r_j} = [0, 0 \cdots, 0, 1]^\mathsf{T} \in \mathbb{R}^{r_j}$, $\mathfrak{B}_{r_j} = [B_{r_j}^\mathsf{T}, 0]^\mathsf{T}$, and $B_{r_j+1} = [0, 0 \cdots, 0, 1]^\mathsf{T} \in \mathbb{R}^{r_j+1}$. Based on Assumption 7.1 and the compact set condition, there exist two positive constants a_2, a_3 such that

$$
\|u_1\| \le a_2 \left[\|A_{0 \sim n-1}\| \left(\|\tilde{X}\| + \|E\| \right) + \frac{\sum_{j \in [m]} \|\delta - \hat{\delta}_j\|}{m} + a_3 \right].
\tag{7.29}
$$

Furthermore, there exist positive constants a_4, a_5, a_6 such that

$$
\max\{\|\varpi_j\|, \|\rho\|\} \le a_4 + a_5 \|\zeta\| + a_6 \|E\|.
\tag{7.30}
$$

Combining (7.24), (7.28)–(7.30), there exist positive constants $a_7, a_8, a_9, a_{10}, a_{11}$ such that

$$
\begin{aligned}
\frac{dV_1}{dt} &< -\ell \kappa_2 \|\zeta\|^2 - \kappa_1 \|e^{(0 \sim n-1)}\|^2 + \ell^{n+1} a_7 \bar{v} \|\zeta\| \\
&\quad + a_8 \|\zeta\| + a_9 \|\zeta\| (a_4 + a_5 \|\zeta\| + a_6 \|e^{(0 \sim n-1)}\|) \\
&\quad + a_{10} \|E\| (a_4 + a_5 \|\zeta\| + a_6 \|E\|) + a_{11} \|\zeta\| \|E\| \\
&= -(\ell \kappa_2 - a_9 a_5) \|\zeta\|^2 + (\ell^{n+1} a_7 \bar{v} + a_8 + a_9 a_4) \|\zeta\| \\
&\quad - (\kappa_1 - a_{10} a_6) \|E\|^2 + a_{10} a_4 \|E\| + (a_9 a_6 + a_{10} a_5 + a_{11}) \|\zeta\| \|E\| \\
&< -\left(\frac{\ell \kappa_2}{2} - a_9 a_5 \right) \|\zeta\|^2 + (\ell^{n+1} a_7 \bar{v} + b_1) \|\zeta\| \\
&\quad - \left(\kappa_1 - a_{10} a_6 - \frac{b_2^2}{2\ell \kappa_2} \right) \|E\|^2 + a_{10} a_4 \|E\|,
\end{aligned}
\tag{7.31}
$$

where $b_1 = a_8 + a_9 a_4$, and $b_2 = a_9 a_6 + a_{10} a_5 + a_{11}$. If the parameter ℓ satisfies

$$\ell > \max \left\{ 1, \frac{2a_9 a_5}{\kappa_2}, \frac{b_2^2}{2\kappa_2(\kappa_1 - a_{10} a_6)} \right\}, \tag{7.32}$$

then the error $[\zeta^\mathsf{T}, E^\mathsf{T}]^\mathsf{T}$ is ultimately uniformly bounded, and the error range can be made sufficiently small by adjusting the parameters ℓ, κ_1, κ_2. Moreover, the ultimate error is affected by the measurement noise, and the main influencing term is $\ell^n \bar{v}$. If the system order is very high, the measurement noise will seriously affect the error system. Section 7.4 will discuss the FTC redesign to suppress the noise. Based on the bounded error $[\zeta^\mathsf{T}, E^\mathsf{T}]^\mathsf{T}$, $\mathcal{V}_j = \zeta_j^\mathsf{T} \mathcal{P}_j \zeta_j$, $j \in [m]$ satisfies

$$\frac{d\mathcal{V}_j}{dt} < -\ell \kappa_2 \|\zeta_j\|^2 + c_1 \|\zeta_j\| + \ell^{r_j+1} c_2 \bar{v} \|\zeta_j\|, \tag{7.33}$$

where c_1, c_2 are positive constants, and the estimation error ζ_j exhibits exponential convergence, i.e.,

$$\|\zeta_j\| \leq \max \left\{ \lambda_j \exp(-\alpha_j \ell t), \frac{\Lambda_j}{\ell} + \ell^{r_j} \beta_j \bar{v} \right\}, \, j \in [m], \tag{7.34}$$

where $\lambda_j, \alpha_j, \Lambda_j, \beta_j$ are positive constants. For $k \in [r_j], j \in [m]$, there exists a positive constant t_1 such that for all $t > t_1$, it holds that

$$\|\xi_{j,k} - \hat{\xi}_{j,k}\| = \ell^{k-1-r_j} \|\zeta_{j,k}\| \leq \ell^{k-2-r_j} \max_{j \in [m]} \Lambda_j + \ell^{k-1} \bar{v} \max_{j \in [m]} \beta_j, \tag{7.35}$$

$$\|\delta - \hat{\delta}_j\| = \|\zeta_{j,r_j+1}\| \leq \ell^{-1} \max_{j \in [m]} \Lambda_j + \ell^{r_j} \bar{v} \max_{j \in [m]} \beta_j. \tag{7.36}$$

Naturally, for $t > t_1$, the actual state observation error and fault estimation error satisfy

$$\|\tilde{X}\| \leq \frac{M_1}{\ell^2} + \ell^{n-1} M_2 \bar{v}, \tag{7.37}$$

$$\|\delta - \hat{\delta}\| \leq \frac{M_3}{\ell} + \ell^n M_4 \bar{v}, \tag{7.38}$$

where M_1, M_2, M_3, M_4 are positive constants. Then, the derivative of $\mathcal{V}_e = E^\mathsf{T} P E, t > t_1$ satisfies

$$\frac{d\mathcal{V}_e}{dt} < -(\kappa_1 - c_3)\|E\|^2 + \left(\frac{c_4}{\ell} + \ell^n c_5 \bar{v} \right) \|E\|, \tag{7.39}$$

where c_3, c_4, c_5 are positive constants. If $\kappa_1 > c_3$, the tracking error E converges, and there exists $t_2 > t_1$ such that

$$\|E\| \leq \frac{M_5}{\ell} + \ell^n M_6 \bar{v}, \forall t > t_2, \tag{7.40}$$

where M_5 and M_6 are positive constants. The proof of Lemma 7.3 is completed.

Remark 7.1 The ultimate error system is mainly influenced by the parameter ℓ, the system order n, and the noise amplitude $\bar{v}$. When the noise amplitude is small, an appropriate ℓ will reduce the error range. To suppress the noise factor term $\ell^n \bar{v}$, two FTC structures will be redesigned in Sect. 7.4.

7.3.2 Multiple Faults: Actuator and Sensor

The FTC structure in Sect. 7.3.1 does not involve a fault diagnosis unit, so this strategy is passive. When the sensor experiences an unbounded fault in amplitude, the aforementioned PFTC cannot achieve system stability through traditional fault compensation. Since the amplitude and rate of sensor faults are not limited, traditional observability cannot guarantee fault tolerance, and the redundant observability in Lemma 7.2 is a sufficient condition for the design of AFTC in this subsection. Similar to the idea in Chap. 6, in order to derive the fault residual signal, (7.9a) is divided into the state observation error when both the actuator and sensor faults exist:

$$\xi - \hat{\xi} = \mathfrak{e} + \mathfrak{r}_{[m]}, \tag{7.41}$$

where $\mathfrak{e}$ represents the error unrelated to the sensor faults, while $\mathfrak{r}_{[m]}$ represents the error solely related to the sensor faults. According to Lemma 7.3, the error $\mathfrak{e}$ satisfies

$$\|\mathfrak{e}\| \leq \gamma(t) \triangleq \sum_{j \in [m]} \max \left\{ \lambda_j \exp(-\alpha_j \ell t), \frac{\Lambda_j}{\ell^2} + \ell^{n-1} \beta_j \bar{v} \right\}. \tag{7.42}$$

If m sensors are in fault-free condition, the error $\mathfrak{r}_{[m]}$ is equal to zero. In the ideal sensor set U, the error $\mathfrak{r}_U$ is also equal to zero. According to Definition 6.3 and Theorem 6.2 in Chap. 6, the following lemma is obtained:

Lemma 7.4 ([55, 56]) *For the nonlinear system (7.1) with at most d ($d < m$) sensor faults, the m measurement outputs y_j can uniquely determine the state $x^{(0 \sim n-1)}$ if and only if the system (7.1) is $2d$-redundant observable.*

When the $2d$-redundant observable system (7.1) has at most d sensor faults, the residual signal based on the sensor set I is designed as

$$\mathfrak{R}_I = \Upsilon_I \left(\text{INV}(\Upsilon_I)(\hat{\xi}_I) \right) - \hat{\xi}_I, \tag{7.43}$$

where $|I| = m - d$, and the definition of $\text{INV}(\Upsilon_I)$ is similar to (7.13).

Lemma 7.5 *If*

$$\|\mathfrak{R}_I\| > \overline{\text{Lip}}(\Upsilon_I \circ \text{INV}(\Upsilon_I) - \mathscr{I})\gamma(t), \tag{7.44}$$

where $\mathscr{I}$ is the identity mapping on the space $\mathbb{R}^{\sum_{j \in I} r_j}$, and $\circ$ represents the operation of mapping multiplication, then there exists a sensor fault in the index set *I*. *Conversely, for a 2d-redundant observable system* (7.1) *with up to d sensor faults, if* (7.44) *does not hold, then the state observer, actuator fault estimator, and sensor fault estimator can be designed as*

$$\hat{X}_I = \text{INV}(\Upsilon_I)(\hat{\xi}_I), \hat{\delta}_I = \frac{\sum_{j \in I} \hat{\delta}_j}{m - d}, \hat{s}_{j,I} = y_j - H_j(\hat{X}_I), j \in [m]. \tag{7.45}$$

At this point, there exist positive constants t_3, M_{I1}, and M_{I2} such that

$$\|x^{(0 \sim n-1)} - \hat{X}_I\| \le \frac{M_{I1}}{\ell^2} + \ell^{n-1} M_{I2} \bar{v}, \forall t > t_3, \tag{7.46}$$

where $\hat{X}_I, \hat{\delta}_I, \hat{s}_{j,I}$ represent the state observation, actuator fault estimation, and sensor fault estimation respectively under the sensor set I.

Proof The proof process of Lemma 7.5 is similar to that of Theorem 6.3.

Based on Lemma 7.5, if

$$\|\mathfrak{R}_{\iota(t)}\| > \overline{\text{Lip}} \left(\Upsilon_{\iota(t)} \circ \text{INV}(\Upsilon_{\iota(t)}) - \mathscr{I} \right) \gamma(t), \tag{7.47}$$

then the dynamic sensor set $\iota(t)$, with $|\iota(t)| = m - d$, is updated to the next set. The definition of $\text{INV}(\Upsilon_{\iota(t)})$ is similar to (7.13). Under m observers (7.11), the AFTC general framework is

$$\begin{cases} \hat{X}_{\iota(t)} = \text{INV}(\Upsilon_{\iota(t)})(\hat{\xi}_{\iota(t)}), \hat{\delta}_{\iota(t)} = \dfrac{\sum_{j \in \iota(t)} \hat{\delta}_j}{m - d}, \hat{s}_{j,\iota(t)} = y_j - H_j(\hat{X}_{\iota(t)}), j \in [m], \\ u_2 = -G^{-1}(\hat{X}_{\iota(t)}) \left[A_{0 \sim n-1} \left(\hat{X}_{\iota(t)} - x_r^{(0 \sim n-1)} \right) + F(\hat{X}_{\iota(t)}) + \hat{\delta}_{\iota(t)} - x_r^{(n)} \right], \end{cases} \tag{7.48}$$

where $\hat{X}_{\iota(t)}, \hat{\delta}_{\iota(t)}, \hat{s}_{j,\iota(t)}$ represent the state estimation, actuator fault estimation, and sensor fault estimation under the sensor set $\iota(t)$, respectively.

Lemma 7.6 *For the nonlinear FAS* (7.1) *with actuator and sensor faults, considering the AFTC general framework* (7.48), *the state estimation error, fault estimation error, and tracking error are ultimately uniformly bounded, i.e., there exist positive constants t_4, t_5, M_7, M_8, M_9, M_{10}, M_{11}, M_{12}, M_{13}, M_{14} such that*

$$\begin{cases} \|x^{(0 \sim n-1)} - \hat{X}_{\iota(t)}\| \le \dfrac{M_7}{\ell^2} + \ell^{n-1} M_8 \bar{v}, \forall t > t_4, \\ \|\delta - \hat{\delta}_{\iota(t)}\| \le \dfrac{M_9}{\ell} + \ell^n M_{10} \bar{v}, \forall t > t_4, \\ \|s_j - \hat{s}_{j,\iota(t)}\| \le \dfrac{M_{11}}{\ell^2} + \ell^{n-1} M_{12} \bar{v}, j \in [m], \forall t > t_4, \\ \|E\| \le \dfrac{M_{13}}{\ell} + \ell^n M_{14} \bar{v}, \forall t > t_5. \end{cases} \tag{7.49}$$

Furthermore, if there is no measurement noise, i.e., $\bar{v} = 0$, then the aforementioned error can converge to a tiny neighborhood around the origin.

Proof Based on Lemma 7.5 and (7.47), $\iota(t)$ will rapidly converge to the ideal sensor set U. Within the compact set, when both actuator and sensor faults occur simultaneously, a Lyapunov function is chosen as

$$V_2 = \sum_{j \in U} \zeta_j^{\mathsf{T}} \mathscr{P}_j \zeta_j + E^{\mathsf{T}} P E. \tag{7.50}$$

The sensor fault estimation error is limited by the state observation error. The remaining proof is similar to Lemma 7.3.

Remark 7.2 Although sensor faults prevent the traditional FTC mechanism from realizing trajectory tracking control, the AFTC mechanism based on analytical redundancy characteristics can yield ideal state observation and trajectory tracking. Its redundant observability stems from the coincidence of each observability subspace. Whether sensor faults are smooth or continuous jumping, the proposed AFTC framework (7.48) can guarantee its FTC performance.

7.4 Advanced Control Redesigns: Saturation and Dead-Zone Methods

7.4.1 Saturation Redesign of Fault-Tolerant Control

The key point of the AFTC general design (7.48) is to determine the fault detection logic (7.47). In (7.42), the noise term $\ell^{n-1}\bar{v}$ in the equation is an important factor affecting the amplitude of $\gamma(t)$, and a too large value of ℓ^{n-1} will reduce the accuracy of fault diagnosis and the effectiveness of the AFTC strategy (7.48). Therefore, in this subsection, the saturation function Sat is utilized to redesign the FTC to reduce the influence of measurement noise.

7.4.1.1 Single Fault: Actuator

In order to reduce the influence of the noise term $\ell^{n-1}\bar{v}$, the following m observers with dynamic saturation functions τ_j, $j \in [m]$ are redesigned as

$$\begin{cases} \dot{\hat{\xi}}'_j = \begin{bmatrix} \hat{\xi}'_{j,2} \\ \vdots \\ \hat{\xi}'_{j,r_j} \\ F_o^j(\hat{\xi}'_j) + \hat{\delta}'_j + G_o^j(\hat{\xi}'_j)u' \end{bmatrix} + \begin{bmatrix} \frac{L_{j,1}}{\ell^{r_j-1}}\mathrm{Sat}_{\tau_j}(\varepsilon'_j) \\ \frac{L_{j,2}}{\ell^{r_j-2}}\mathrm{Sat}_{\tau_j}(\varepsilon'_j) \\ \vdots \\ L_{j,r_j}\mathrm{Sat}_{\tau_j}(\varepsilon'_j) \end{bmatrix}, \\ \dot{\hat{\delta}}'_j = \ell L_{j,r_j+1}\mathrm{Sat}_{\tau_j}(\varepsilon'_j), \\ \dot{\tau}_j = -\ell\theta\tau_j + \ell\mu\|\varepsilon'_j\|, \, \tau_j \geq 0, \end{cases} \tag{7.51}$$

where θ and μ are positive constants satisfying $\mu > \theta$, and $\hat{\xi}'_j = \left[\hat{\xi}'_{j,1}, \hat{\xi}'_{j,2}, \cdots, \hat{\xi}'_{j,r_j}\right]^{\mathsf{T}}$ represents the state observation. $\varepsilon'_j = \ell^{r_j}(y_j - \hat{\xi}'_{j,1})$, and $\hat{\delta}'_j$ is the actuator fault estimation under the jth measurement. The overall observation vector is $\hat{\xi}' = \left[\hat{\xi}'^{\mathsf{T}}_1, \hat{\xi}'^{\mathsf{T}}_2, \cdots, \hat{\xi}'^{\mathsf{T}}_m\right]$, and the corresponding state estimation $\hat{X}'$ is

$$\hat{X}' = \mathrm{INV}(\Upsilon_{[m]})(\hat{\xi}') \triangleq \mathrm{Sat}_{\Xi_x}\left(\Upsilon^{-1}_{[m]}\left([\mathrm{Sat}_{\Xi_1}\hat{\xi}'_1, \mathrm{Sat}_{\Xi_2}\hat{\xi}'_2, \cdots, \mathrm{Sat}_{\Xi_m}\hat{\xi}'_m]^{\mathsf{T}}\right)\right). \tag{7.52}$$

The fusion estimation of the actuator fault is

$$\hat{\delta}' = \frac{\sum_{j\in[m]} \hat{\delta}'_j}{m}. \tag{7.53}$$

Therefore, under the saturation strategy, an FTC framework is redesigned as

$$u'_1 = -G^{-1}(\hat{X}')\left[A_{0\sim n-1}\left(\hat{X}' - x_r^{(0\sim n-1)}\right) + F(\hat{X}') + \hat{\delta}' - x_r^{(n)}\right]. \tag{7.54}$$

Theorem 7.1 *For the nonlinear FAS (7.1) with only the actuator fault, considering the FTC saturation design (7.51)–(7.54), the state estimation error, actuator fault estimation error and tracking error are ultimately uniformly bounded, i.e., there exist positive constants $t'_1, t'_2, M'_1, M'_2, M'_3, M'_4, M'_5, M'_6$ such that*

$$\begin{cases} \|x^{(0\sim n-1)} - \hat{X}'\| \leq \dfrac{M'_1}{\ell^2} + \ell^{n-1}M'_2\bar{v}, \forall t > t'_1, \\[2mm] \|\delta - \hat{\delta}'\| \leq \dfrac{M'_3}{\ell} + \ell^n M'_4\bar{v}, \forall t > t'_1, \\[2mm] \|E\| \leq \dfrac{M'_5}{\ell} + \ell^n M'_6\bar{v}, \forall t > t'_2. \end{cases} \tag{7.55}$$

Furthermore, if there is no measurement noise, i.e., $\bar{v} = 0$, then the aforementioned error can converge to a tiny neighborhood around the origin.

Proof For each measurement y_j, the corresponding estimation error is

$$\zeta'_j = [\zeta'_{j,1}, \zeta'_{j,2}, \cdots, \zeta'_{j,r_j}, \zeta'_{j,r_j+1}]^{\mathsf{T}}$$
$$= [\ell^{r_j}(\xi_{j,1} - \hat{\xi}'_{j,1}), \ell^{r_j-1}(\xi_{j,2} - \hat{\xi}'_{j,2}), \cdots, \delta - \hat{\delta}'_j]^{\mathsf{T}}, \tag{7.56}$$

The derivatives of ζ_j' and τ_j are calculated as

$$
\begin{cases}
\dot{\zeta}_{j,k}' = \ell\zeta_{j,k+1}' - \ell L_{j,k}\zeta_{j,1}' + \ell L_{j,k}q_j + \ell L_{j,k}Q_j, \\
\dot{\zeta}_{j,r_j}' = \ell\zeta_{j,r_j+1}' - \ell L_{j,r_j}\zeta_{j,1}' + \ell L_{j,r_j}q_j + \ell L_{j,r_j}Q_j + \varpi_j', \\
\dot{\zeta}_{j,r_j+1}' = \dot{\delta} - \ell L_{j,r_j+1}\zeta_{j,1}' + \ell L_{j,r_j+1}q_j + \ell L_{j,r_j+1}Q_j, \\
\dot{\tau}_j = -\ell\theta\tau_j + \ell\mu\|\zeta_{j,1}' + \ell^{r_j}v_j\|,
\end{cases}
\tag{7.57}
$$

wher $j \in [m], k \in [r_j - 1]$,

$$
q_j = \zeta_{j,1}' - \mathrm{Sat}_{\tau_j}(\zeta_{j,1}'),
\tag{7.58}
$$

$$
Q_j = \mathrm{Sat}_{\tau_j}(\zeta_{j,1}') - \mathrm{Sat}_{\tau_j}(\zeta_{j,1}' + \ell^{r_j}v_j),
\tag{7.59}
$$

$$
\varpi_j' = \ell[F_o^j(\xi_j) - F_o^j(\hat{\xi}_j') + (G_o^j(\xi_j) - G_o^j(\hat{\xi}_j'))u_1'].
\tag{7.60}
$$

The dynamic equations of high-order tracking error E are

$$
\begin{cases}
\dot{e}_i = e_{i+1}, i \in [n - 1], \\
\dot{e}_n = \rho' + A_{0\sim n-1}\tilde{X}' - A_{0\sim n-1}E,
\end{cases}
\tag{7.61}
$$

where

$$
\tilde{X}' = x^{(0\sim n-1)} - \hat{X}',
\tag{7.62}
$$

$$
\rho' = F(x^{(0\sim n-1)}) - F(\hat{X}') + \delta - \hat{\delta}' + [G(x^{(0\sim n-1)}) - G(\hat{X}')]u_1'.
\tag{7.63}
$$

After certain mathematical derivations, it follows that

$$
\begin{cases}
\|q_j\| \le \|\zeta_{j,1}'\|, \|Q_j\| \le \|\ell^{r_j}v_j\| \le \ell^{r_j}\bar{v}, \\
\max\left\{\|\varpi_j'\|, \|\rho'\|\right\} \le a_1' + a_2'\|\zeta'\| + a_3'\|E\|,
\end{cases}
\tag{7.64}
$$

where a_1', a_2', a_3' are positive constants.

A Lyapunov function is chosen as

$$
V_1' = \sum_{j=1}^{m} \zeta_j'^{\mathsf{T}}\mathscr{P}_j\zeta_j' + v_j\tau_j + \varkappa_j \max\left\{\|\zeta_{j,1}'\| - \tau_j, 0\right\} + E^{\mathsf{T}}PE,
\tag{7.65}
$$

where $\varkappa_j$ and v_j are positive constants satisfying $\varkappa_j > v_j$. For the locally Lipschitz function V_1', the Dini derivative is defined as

$$
\mathrm{D}^+V_1' \triangleq \frac{\lim_{h\to 0^+}(V_1'(t + h) - V_1'(t))}{h}.
\tag{7.66}
$$

Additionally, define

$$\mathfrak{V}_j = \zeta_j'^{\mathsf{T}} \mathscr{P}_j \zeta_j' + v_j \tau_j + \varkappa_j \max\{\|\zeta_{j,1}'\| - \tau_j, 0\}, \tag{7.67}$$

then $V_1' = \sum_{j=1}^m \mathfrak{V}_j + E^{\mathsf{T}} P E$, and the Dini derivative of $\mathfrak{V}_j$ can be discussed in the following two cases:

Case 1: $\|\zeta_{j,1}'\| < \tau_j$. In this case, $q_j = 0$, $\mathfrak{V}_j = \zeta_j'^{\mathsf{T}} \mathscr{P}_j \zeta_j' + v_j \tau_j$. At this time, the Dini derivative of $\mathfrak{V}_j$ is

$$\begin{aligned}
\mathrm{D}^+ \mathfrak{V}_j &= \ell \zeta_j'^{\mathsf{T}} (\mathscr{P}_j \Psi_j + \Psi_j^{\mathsf{T}} \mathscr{P}_j) \zeta_j' \\
&\quad + 2\ell Q_j L_j \mathscr{P}_j \zeta_j' + 2\dot{\delta} B_{r_j+1}^{\mathsf{T}} \mathscr{P}_j \zeta_j' + 2\varpi_j' \mathfrak{B}_{r_j}^{\mathsf{T}} \mathscr{P}_j \zeta_j' \\
&\quad + v_j \left[-\ell \theta \tau_j + \ell \mu \|\zeta_{j,1}'\| + \ell^{r_j} v_j \| \right].
\end{aligned} \tag{7.68}$$

Combining (7.26) and (7.64), there exist positive constants a_4' and a_5' such that

$$\begin{aligned}
\mathrm{D}^+ \mathfrak{V}_j &< -\ell \kappa_2 \|\zeta_j'\|^2 + (\ell^{r_j+1} a_4' \bar{v} + \ell v_j \mu) \|\zeta_j'\| \\
&\quad + a_5' \|\zeta_j'\|(a_1' + a_2' \|\zeta'\| + a_3' \|E\|) - \ell v_j \theta \tau_j + \ell^{r_j+1} v_j \mu \bar{v} \\
&< -\left(\frac{\ell \kappa_2}{2} - a_2' a_5' \right) \|\zeta_j'\|^2 + (\ell^{r_j+1} a_4' \bar{v} + a_1' a_5' + \ell v_j \mu) \|\zeta_j'\| \\
&\quad + \frac{a_3'^2 a_5'^2}{2\ell \kappa_2} \|E\|^2 + \ell^{r_j+1} v_j \mu \bar{v}.
\end{aligned} \tag{7.69}$$

Case 2: $\|\zeta_{j,1}'\| > \tau_j$. In this case, $q_j \neq 0$, $\mathfrak{V}_j = \zeta_j'^{\mathsf{T}} \mathscr{P}_j \zeta_j' - (\varkappa_j - v_j)\tau_j + \varkappa_j \|\zeta_{j,1}'\|$. At this time, the Dini derivative of $\mathfrak{V}_j$ is

$$\begin{aligned}
\mathrm{D}^+ \mathfrak{V}_j &= \ell \zeta_j'^{\mathsf{T}} (\mathscr{P}_j \Psi_j + \Psi_j^{\mathsf{T}} \mathscr{P}_j) \zeta_j' + 2\ell(q_j + Q_j) L_j \mathscr{P}_j \zeta_j' + 2\dot{\delta} B_{r_j+1}^{\mathsf{T}} \mathscr{P}_j \zeta_j' \\
&\quad + 2\varpi_j' \mathfrak{B}_{r_j}^{\mathsf{T}} \mathscr{P}_j \zeta_j' + (\varkappa_j - v_j) \left[\ell \theta \tau_j - \ell \mu \|\zeta_{j,1}'\| + \ell^{r_j} v_j \| \right] \\
&\quad + \varkappa_j \left\| \ell \zeta_{j,2}' - \ell L_{j,1} \zeta_{j,1}' + \ell L_{j,1} q_1 + \ell L_{j,1} Q_1 \right\|.
\end{aligned} \tag{7.70}$$

Similarly, there exist positive constants a_6' and a_7' such that

$$\begin{aligned}
\mathrm{D}^+ \mathfrak{V}_j &< -\ell \left[\kappa_2 - 2\|L_j \mathscr{P}_j\| + \varkappa_j (L_{j,1} - 2) \right] \|\zeta_j'\|^2 + \ell^{r_j+1} a_6' \bar{v} \|\zeta_j'\| \\
&\quad + a_7' \|\zeta_j'\|(a_1' + a_2' \|\zeta'\| + a_3' \|E\|) + \ell^{r_j+1} (\varkappa_j - v_j) \mu \bar{v} \\
&< -\ell \left[\frac{\kappa_2}{2} - 2\|L_j \mathscr{P}_j\| + \varkappa_j (L_{j,1} - 2) - \frac{a_2' a_7'}{\ell} \right] \|\zeta_j'\|^2
\end{aligned}$$

$$+ (\ell^{r_j+1} a_6' \bar{v} + a_1' a_7') \|\zeta_j'\| + \frac{a_3'^2 a_7'^2}{2\ell \kappa_2} \|E\|^2 + \ell^{r_j+1} (x_j - v_j) \mu \bar{v}.$$

$$(7.71)$$

Define

$$\aleph_{j,0} = \min \left\{ \frac{\kappa_2}{2} - \frac{a_2' a_5'}{\ell}, \frac{\kappa_2}{2} - 2\|L_j \mathscr{P}_j\| + x_j (L_{j,1} - 2) - \frac{a_2' a_7'}{\ell} \right\}, \qquad (7.72)$$

there exist postive constants $\aleph_{j,1}$, $\aleph_{j,2}$, $\aleph_{j,3}$, where $j \in [m]$, such that the following equation holds:

$$D^+ \mathfrak{V}_j < -\ell \aleph_{j,0} \|\zeta_j'\|^2 + \ell^{r_j+1} \aleph_{j,1} \bar{v} \|\zeta_j'\| + \aleph_{j,2} \|E\|^2 + \ell^{r_j+1} \aleph_{j,3} \bar{v}. \qquad (7.73)$$

The Dini derivative of V_1' is calculated as

$$D^+ V_1' = \sum_{j=1}^{m} D^+ \mathfrak{V}_j + E^{\mathsf{T}} (P \Phi_A + \Phi_A^{\mathsf{T}} P) E + 2\rho' B_n^{\mathsf{T}} P E + 2A_{0 \sim n-1} \tilde{X}' B_n^{\mathsf{T}} P E$$

$$= \sum_{j=1}^{m} D^+ \mathfrak{V}_j - \kappa_1 \|E\|^2 + 2\rho' B_n^{\mathsf{T}} P E + 2A_{0 \sim n-1} \tilde{X}' B_n^{\mathsf{T}} P E. \qquad (7.74)$$

Since $\text{INV}(\Upsilon_{[m]})$ is Lipschitz, it follows that

$$\|\tilde{X}'\| = \|\text{INV}(\Upsilon_{[m]})(\xi) - \text{INV}(\Upsilon_{[m]})(\hat{\xi}')\| \le a_8' \|\zeta'\|, \qquad (7.75)$$

where a_8' is a positive constant. Then, the Dini derivative of V_1' satisfies

$$D^+ V_1' < -\ell \min_{j \in [m]} \aleph_{j,0} \|\zeta'\|^2 + \ell^{r_j+1} \max_{j \in [m]} \aleph_{j,1} \bar{v} \|\zeta'\|$$

$$+ m \ell^{r_j+1} \max_{j \in [m]} \aleph_{j,3} \bar{v} + m \max_{j \in [m]} \aleph_{j,2} \|E\|^2$$

$$- \kappa_1 \|E\|^2 + 2a_8' \|A_{0 \sim n-1}\| \|B_n^{\mathsf{T}} P\| \|\zeta'\| \|E\|$$

$$+ 2(a_1' + a_2' \|\zeta'\| + a_3' \|E\|) \|B_n^{\mathsf{T}} P\| \|E\|$$

$$< -\ell (\min_{j \in [m]} \aleph_{j,0} - \frac{1}{4}) \|\zeta'\|^2 + \ell^{n+1} \max_{j \in [m]} \aleph_{j,1} \bar{v} \|\zeta'\|$$

$$- (\kappa_1 - b_3') \|E\|^2 + b_4' \|E\| + m \ell^{n+1} \max_{j \in [m]} \aleph_{j,3} \bar{v}, \qquad (7.76)$$

where

$$b_1' = 2a_8' \|A_{0 \sim n-1}\| \|B_n^{\mathsf{T}} P\|, \ b_2' = 2a_2' \|B_n^{\mathsf{T}} P\|, \qquad (7.77)$$

$$b_3' = \frac{2b_1'^2 + 2b_2'^2}{\ell} + m \max_{j \in [m]} \aleph_{j,2} + 2a_3' \| B_n^\mathsf{T} P \|, \tag{7.78}$$

$$b_4' = 2a_1' \| B_n^\mathsf{T} P \|. \tag{7.79}$$

If the parameters satisfy

$$\begin{cases} \min_{j \in [m]} \aleph_{j,0} - \dfrac{1}{4} > 0, \\ \kappa_1 - b_3' > 0, \end{cases} \tag{7.80}$$

The overall error $[\zeta'^\mathsf{T}, E^\mathsf{T}]^\mathsf{T}$ is ultimately uniformly bounded, and the error range can be made sufficiently small by adjusting the parameters ℓ, κ_1, and κ_2. However, the performance of the proposed FTC is still affected by the noise term $\ell^n \bar{v}$.

The following analysis can be referred to (7.34)–(7.40) and Lemma 7.3. The error ζ_j' still converges exponentially, i.e.,

$$\| \zeta_j' \| \leq \max \left\{ \lambda_j' \exp(-\alpha_j' \ell t), \frac{\Lambda_j'}{\ell} + \ell^{r_j} \beta_j' \bar{v} \right\}, \ j \in [m], \tag{7.81}$$

where $\lambda_j', \alpha_j', \Lambda_j', \beta_j', j \in [m]$ are positive constants. Furthermore, there exist positive constants t_1' and $t_2', t_2' > t_1'$ such that the actual state observation error, fault estimation error, and tracking error satisfy

$$\begin{cases} \| \tilde{X}' \| \leq \dfrac{M_1'}{\ell^2} + \ell^{n-1} M_2' \bar{v}, \forall t > t_1', \\ \| \delta - \hat{\delta}' \| \leq \dfrac{M_3'}{\ell} + \ell^n M_4' \bar{v}, \forall t > t_1', \\ \| E \| \leq \dfrac{M_5'}{\ell} + \ell^n M_6' \bar{v}, \forall t > t_2', \end{cases}$$

where $M_1', M_2', M_3', M_4', M_5', M_6'$ are positive constants. Theorem 7.1 has been proved.

Remark 7.3 Since the measurement noise cannot be completely eliminated, the stability conclusions under the two controllers (7.15) and (7.54) are consistent. Although the physical meanings of Theorem 7.1 and Lemma 7.3 are the same, the FTC saturation design has better suppression performance for the noise term $\ell^n \bar{v}$. The error feedback term $\mathrm{Sat}_{\tau_j}(\ell^{r_j}(\xi_{j,1} + v_j - \hat{\xi}_{j,1}'))$ in (7.51) adopts a saturation suppression strategy, so the large error caused by noise will be quickly limited. This strategy is, in a sense, similar to a filter in the frequency domain. Due to its nonlinear characteristics, it is difficult to conduct a performance comparison analysis of the two structures. Therefore, a theoretical comparison analysis under a linear system is given later.

7.4.1.2 Multiple Faults: Actuator and Sensor

Under the saturation strategy, the state observation error of (7.9) with actuator and sensor faults is

$$\xi - \hat{\xi}' = \mathfrak{e}' + \mathfrak{r}'_{[m]}, \tag{7.82}$$

where $\mathfrak{e}'$ is an error unrelated to the sensor faults, while $\mathfrak{r}'_{[m]}$ is only related to the sensor faults. According to Theorem 7.1, it follows that

$$\|\mathfrak{e}'\| \leq \gamma'(t) \triangleq \sum_{j \in [m]} \max \left\{ \lambda'_j \exp(-\alpha'_j \ell t), \frac{\Lambda'_j}{\ell^2} + \ell^{n-1} \beta'_j \bar{v} \right\}. \tag{7.83}$$

If all m sensors are in fault-free condition, then the error $\mathfrak{r}'_{[m]}$ is equal to zero. In the ideal sensor set U, the error $\mathfrak{r}'_U$ is also equal to zero. For a $2d$-redundant observable system with at most d sensor faults, the following fault residual signal based on the sensor set I, $|I| = m - d$ is designed as

$$\mathfrak{R}'_I = \Upsilon_I \left(\text{INV}(\Upsilon_I)(\hat{\xi}'_I) \right) - \hat{\xi}'_I. \tag{7.84}$$

Theorem 7.2 *If*

$$\|\mathfrak{R}'_I\| > \overline{\text{Lip}}(\Upsilon_I \circ \text{INV}(\Upsilon_I) - \mathscr{I})\gamma'(t), \tag{7.85}$$

then there exists a sensor fault in the index set I. Conversely, if for the 2d redundant observable system (7.1) with at most d sensor faults, (7.85) does not hold. the state observer, actuator fault estimator, and sensor fault estimator are designed as

$$\hat{X}'_I = \text{INV}(\Upsilon_I)(\hat{\xi}'_I), \hat{\delta}'_I = \frac{\sum_{j \in I} \hat{\delta}'_j}{m - d}, \hat{s}'_{j,I} = y_j - H_j(\hat{X}'_I), j \in [m]. \tag{7.86}$$

At this point, there exist positive constants t'_3, M'_{I1}, and M'_{I2} such that

$$\|x^{(0 \sim n-1)} - \hat{X}'_I\| \leq \frac{M'_{I1}}{\ell^2} + \ell^{n-1} M'_{I2} \bar{v}, \forall t > t'_3, \tag{7.87}$$

where $\hat{X}'_I$, $\hat{\delta}'_I$, and $\hat{s}'_{j,I}$ represent the state observation, actuator fault estimation, and sensor fault estimation respectively for the sensor set I.

Proof The proof process is similar to Lemma 7.5.

If

$$\|\mathfrak{R}'_{\iota(t)}\| > \overline{\text{Lip}}(\Upsilon_{\iota(t)} \circ \text{INV}(\Upsilon_{\iota(t)}) - \mathscr{I})\gamma'(t). \tag{7.88}$$

then the dynamic sensor set $\iota(t)$, with $|\iota(t)| = m - d$, is updated to the next set $\iota(t^+)$. The AFTC saturation design is

$$
\begin{cases}
\hat{X}'_{\iota(t)} = \text{INV}(\Upsilon_{\iota(t)})(\hat{\xi}'_{\iota(t)}), \hat{\delta}'_{\iota(t)} = \dfrac{\sum_{j \in \iota(t)} \hat{\delta}'_j}{m - d}, \hat{s}'_{j,\iota(t)} = y_j - H_j(\hat{X}'_{\iota(t)}), j \in [m], \\
u'_2 = -G^{-1}(\hat{X}'_{\iota(t)}) \left[A_{0 \sim n-1} \left(\hat{X}'_{\iota(t)} - x_r^{(0 \sim n-1)} \right) + F(\hat{X}'_{\iota(t)}) + \hat{\delta}'_{\iota(t)} - x_r^{(n)} \right],
\end{cases}
\tag{7.89}
$$

where $\hat{X}'_{\iota(t)}, \hat{\delta}'_{\iota(t)}, \hat{s}'_{j,\iota(t)}$ are the state observation, actuator fault estimation, and sensor fault estimation under the sensor set $\iota(t)$, respectively.

Theorem 7.3 *For the nonlinear FAS (7.1) with actuator and sensor faults, considering the AFTC saturation design (7.89), the state estimation error, fault estimation error, and tracking error are ultimately uniformly bounded, i.e., there exist positive constants $t'_4, t'_5, M'_7, M'_8, M'_9, M'_{10}, M'_{11}, M'_{12}, M'_{13}, M'_{14}$ such that*

$$
\begin{cases}
\| x^{(0 \sim n-1)} - \hat{X}'_{\iota(t)} \| \leq \dfrac{M'_7}{\ell^2} + \ell^{n-1} M'_8 \bar{v}, \forall t > t'_4, \\
\| \delta - \hat{\delta}'_{\iota(t)} \| \leq \dfrac{M'_9}{\ell} + \ell^n M'_{10} \bar{v}, \forall t > t'_4, \\
\| s_j - \hat{s}'_{j,\iota(t)} \| \leq \dfrac{M'_{11}}{\ell^2} + \ell^{n-1} M'_{12} \bar{v}, j \in [m], \forall t > t'_4, \\
\| E \| \leq \dfrac{M'_{13}}{\ell} + \ell^n M'_{14} \bar{v}, \forall t > t'_5.
\end{cases}
\tag{7.90}
$$

Furthermore, if there is no measurement noise, i.e., $\bar{v} = 0$, then the aforementioned error can converge to a tiny neighborhood around the origin.

Proof Based on Theorem 7.2 and (7.88), $\iota(t)$ will rapidly converge to the ideal sensor set U. Within the compact set, when the system simultaneously experiences actuator and sensor faults, a Lyapunov function is chosen as

$$
V'_2 = \sum_{j \in I} \zeta_j^\mathsf{T} \mathscr{P}_j \zeta_j + v_j \tau_j + \varkappa_j \max \left\{ \| \zeta'_{j,1} \| - \tau_j, 0 \right\} + E^\mathsf{T} P E.
\tag{7.91}
$$

The sensor estimation error is limited by the state observation error. The remaining proof is similar to Theorem 7.1.

7.4.1.3   Performance Comparison Analysis of Linear FASs

Although it is difficult to conduct a noise suppression comparison analysis between AFTC saturation design and AFTC general design in the nonlinear FAS model, certain noise suppression performance comparisons can be made in the linear FAS model. The fault detection conditions (7.44) and (7.85) are the key to both AFTC designs, while the observation performance is the core of the two fault detection conditions. Moreover, the influence of the noise term $\ell^n \bar{v}$ on the observation performance mainly occurs in the case of sensorless faults. Therefore, the comparison of system observation error to some extent reflects the superiority of saturation design.

Consider the following a simple linear FAS:

$$\begin{cases} x^{(n)} = \mathscr{A} x^{(0 \sim n-1)} + \mathscr{B} u + \delta, \\ y_1 = x + v_1, \end{cases} \tag{7.92}$$

where $x, u, y \in \mathbb{R}$ are the systems state, the control input, and the measurement output, respectively. $\mathscr{A} = [\mathscr{A}_1, \mathscr{A}_2, \cdots, \mathscr{A}_n] \in \mathbb{R}^{1 \times n}$ and $\mathscr{B} \in \mathbb{R}$ are known, v_1 is noise, and δ is assumed to be a constant fault.

The observer in the general design is

$$\begin{cases} \dot{\hat{X}} = \begin{bmatrix} \hat{x}_2 \\ \vdots \\ \hat{x}_n \\ \mathscr{A}\hat{X} + \hat{\delta} \end{bmatrix} + \begin{bmatrix} \ell \mathfrak{L}_1 \\ \ell^2 \mathfrak{L}_1 \\ \vdots \\ \ell^n \mathfrak{L}_n \end{bmatrix} (y_1 - \hat{x}_1) + \begin{bmatrix} 0 \\ \vdots \\ 0 \\ \mathscr{B} \end{bmatrix} u, \\ \dot{\hat{\delta}} = \ell^{n+1} \mathfrak{L}_{n+1} (y_1 - \hat{x}_1), \end{cases} \tag{7.93}$$

where $\mathfrak{L} = [\mathfrak{L}_1, \mathfrak{L}_2, \cdots, \mathfrak{L}_{n+1}]^\mathsf{T}$ should shape a Hurwitz matrix

$$\Psi = \begin{bmatrix} -\mathfrak{L}_1 & 1 & 0 & \cdots & 0 \\ -\mathfrak{L}_2 & 0 & 1 & \cdots & 0 \\ \vdots & & & \ddots & \\ -\mathfrak{L}_n & 0 & 0 & \cdots & 1 \\ -\mathfrak{L}_{n+1} & 0 & 0 & \cdots & 0 \end{bmatrix} \in \mathbb{R}^{(n+1) \times (n+1)}. \tag{7.94}$$

Define the estimation error

$$\mathfrak{X} = [\mathfrak{x}_1, \mathfrak{x}_2, \cdots, \mathfrak{x}_n, \mathfrak{x}_{n+1}]^\mathsf{T} = [\ell^n(x_1 - \hat{x}_1), \ell^{n-1}(x_2 - \hat{x}_2), \cdots, \delta - \hat{\delta}]^\mathsf{T}, \tag{7.95}$$

and the derivative of $\mathfrak{X}$ is

$$\begin{cases} \dot{\mathfrak{x}}_k = \ell \mathfrak{x}_{k+1} - \ell \mathfrak{L}_k \mathfrak{x}_1 - \ell^{n+1} \mathfrak{L}_k v_1, \, k \in [n-1], \\ \dot{\mathfrak{x}}_n = \ell \mathfrak{x}_{n+1} - \ell \mathfrak{L}_n \mathfrak{x}_1 - \ell^{n+1} \mathfrak{L}_n v_1 + \varpi_{\mathfrak{x}}, \\ \dot{\mathfrak{x}}_{n+1} = -\ell \mathfrak{L}_{n+1} \mathfrak{x}_1 - \ell^{n+1} \mathfrak{L}_{n+1} v_1, \end{cases} \tag{7.96}$$

where

$$\varpi_{\mathfrak{x}} = \ell \mathscr{A} \left[\frac{1}{\ell^n} \mathfrak{x}_1, \frac{1}{\ell^{n-1}} \mathfrak{x}_2, \cdots, \frac{1}{\ell} \mathfrak{x}_n \right]^{\mathsf{T}}. \tag{7.97}$$

The error dynamics (7.96) can be rewritten in the following matrix form

$$\dot{\mathfrak{x}} = \ell \Psi_{\mathfrak{x}} \mathfrak{x} - \ell^{n+1} \mathfrak{L} v_1, \tag{7.98}$$

where

$$\Psi_{\mathfrak{x}} = \begin{bmatrix} -\mathfrak{L}_1 & 1 & 0 & \cdots & 0 \\ -\mathfrak{L}_2 & 0 & 1 & \cdots & 0 \\ \vdots & & & \ddots & \\ -\mathfrak{L}_n + \frac{\mathscr{A}_1}{\ell^n} & \frac{\mathscr{A}_2}{\ell^{n-1}} & \frac{\mathscr{A}_3}{\ell^{n-2}} & \cdots & 1 \\ -\mathfrak{L}_{n+1} & 0 & 0 & \cdots & 0 \end{bmatrix}. \tag{7.99}$$

Correspondingly, the observer in the saturation design is

$$\begin{cases} \dot{\hat{X}}' = \begin{bmatrix} \hat{x}_2' \\ \vdots \\ \hat{x}_n' \\ \mathscr{A}\hat{X}' + \hat{\delta}' + \mathscr{B}u \end{bmatrix} + \begin{bmatrix} \frac{\mathfrak{L}_1}{\ell^{n-1}} \operatorname{Sat}_{\tau_1}(\varepsilon') \\ \frac{\mathfrak{L}_2}{\ell^{n-2}} \operatorname{Sat}_{\tau_1}(\varepsilon') \\ \vdots \\ \mathfrak{L}_n \operatorname{Sat}_{\tau_1}(\varepsilon') \end{bmatrix}, \\ \dot{\hat{\delta}}' = \ell \mathfrak{L}_{n+1} \operatorname{Sat}_{\tau_1}(\varepsilon'), \\ \dot{\tau}_1 = -\ell \theta \tau_1 + \ell \mu \|\varepsilon'\|, \tau_1 \geq 0, \end{cases} \tag{7.100}$$

where $\varepsilon' = \ell^n (y - \hat{x}_1')$, and τ_1 is a dynamic saturation bound. Define the observation error

$$\mathfrak{X}' = [\mathfrak{x}_1', \mathfrak{x}_2', \cdots, \mathfrak{x}_n', \mathfrak{x}_{n+1}']^{\mathsf{T}} = [\ell^n (x_1 - \hat{x}_1'), \ell^{n-1}(x_2 - \hat{x}_2'), \cdots, \delta - \hat{\delta}']^{\mathsf{T}}, \tag{7.101}$$

and the derivative of $\mathfrak{X}'$ is

$$\begin{cases} \dot{\mathfrak{x}}_k' = \ell \mathfrak{x}_{k+1}' - \ell \mathfrak{L}_k \operatorname{Sat}_{\tau_1}(\mathfrak{x}_1' + \ell^n v_1), k \in [n-1], \\ \dot{\mathfrak{x}}_n' = \ell \mathfrak{x}_{n+1}' - \ell \mathfrak{L}_n \operatorname{Sat}_{\tau_1}(\mathfrak{x}_1' + \ell^n v_1) + \varpi_{\mathfrak{x}}', \\ \dot{\mathfrak{x}}_{n+1}' = -\ell \mathfrak{L}_{n+1} \operatorname{Sat}_{\tau_1}(\mathfrak{x}_1' + \ell^n v_1), \\ \dot{\tau}_1 = -\ell \theta \tau_1 + \ell \mu \|\mathfrak{x}_1' + \ell^n v_1\|, \end{cases} \tag{7.102}$$

where

$$\varpi_{\mathfrak{x}}' = \ell \mathscr{A} [\frac{1}{\ell^n} \mathfrak{x}_1', \frac{1}{\ell^{n-1}} \mathfrak{x}_2', \cdots, \frac{1}{\ell} \mathfrak{x}_n']^{\mathsf{T}}. \tag{7.103}$$

The error dynamics (7.102) can be rewritten in the following matrix form

$$
\begin{cases}
\dot{\mathfrak{x}}' = \ell \Psi'_{\mathfrak{x}} \mathfrak{x}' - \ell \mathcal{L} \mathrm{Sat}_{\tau_1}(\mathfrak{x}'_1 + \ell^n v_1), \\
\dot{\tau}_1 = -\ell \theta \tau_1 + \ell \mu \|\mathfrak{x}'_1 + \ell^n v_1\|,
\end{cases}
\tag{7.104}
$$

where

$$
\Psi'_{\mathfrak{x}} = \begin{bmatrix}
0 & 1 & 0 & \cdots & 0 \\
0 & 0 & 1 & \cdots & 0 \\
\vdots & & & \ddots & \\
\frac{\mathcal{A}_1}{\ell^n} & \frac{\mathcal{A}_2}{\ell^{n-1}} & \frac{\mathcal{A}_3}{\ell^{n-2}} & \cdots & 1 \\
0 & 0 & 0 & \cdots & 0
\end{bmatrix}.
\tag{7.105}
$$

In order to compare the noise suppression performance of the two observers (7.93) and (7.100), the noise is selected as

$$
v_1 = \mathscr{D}_p(t) \triangleq \begin{cases}
\dfrac{1}{p}, & 0 < t \le p, \\
0, & t \ge p,
\end{cases}
\tag{7.106}
$$

where $\mathscr{D}_p(t)$, $p \to 0^+$ represents the impulse function. Then, $\|\mathfrak{X}'(p)\|/\|\mathfrak{X}(p)\|$, $p \to 0^+$ is used as a performance comparison metric.

Theorem 7.4 *For the linear FAS models (7.92) under the two observers (7.93) and (7.100), if $\mathfrak{e}'(0) = \mathfrak{e}(0)$, $\tau_1(0) = 0$, $v_1 = \mathscr{D}_p(t)$ and $\Psi_{\mathfrak{x}}$ is invertible, then the following conclusion holds:*

$$
\frac{\|\mathfrak{X}'(p)\|}{\|\mathfrak{X}(p)\|} \le p\ell\mu \frac{1 + o(p)}{1 - o(p)}, p \to 0^+,
\tag{7.107}
$$

were, $o(p)$ represents an infinitesimal quantity.

Proof Based on the error dynamics (7.98), it can be obtained that

$$
\mathfrak{X}(p) = -\int_0^p \exp(\ell \Psi_{\mathfrak{x}}(p - t))\ell^{n+1} \mathcal{L} \mathscr{D}_p(t) dt
$$

$$
= -\ell^n \left(\exp(\ell \Psi_{\mathfrak{x}} p) - I_{n+1} \right) \Psi_{\mathfrak{x}}^{-1} \frac{\mathcal{L}}{p}.
\tag{7.108}
$$

According to the Taylor expansion

$$
\exp(\ell \Psi_{\mathfrak{x}} p) = I_{n+1} + \ell \Psi_{\mathfrak{x}} p + o(p^2), p \to 0^+,
\tag{7.109}
$$

where the infinitesimal quantity $o(p^2)$ could be a matrix or a scalar, and it follows that

$$\|\mathfrak{X}(p)\| = \frac{\ell^{n+1}\|\mathfrak{L}\|}{p}(p + o(p^2)) \geq \ell^{n+1}\|\mathfrak{L}\|(1 - o(p)), \, p \to 0^+. \tag{7.110}$$

According to the error dynamics (7.104), it turns out that

$$\mathfrak{X}'(p) = \int_0^p \exp(\ell\Psi_{\mathfrak{x}}'(p - t_p))\ell\mathfrak{L}\mathfrak{f}(t_p)\mathrm{d}t_p, \tag{7.111}$$

where

$$\begin{aligned}
\mathfrak{f}(t_p) &\leq \int_0^{t_p} \exp(-\ell\theta(t_p - t))\ell\mu(\mathfrak{x}_1' + \ell^n v_0)\mathrm{d}t \\
&\leq \frac{\mu}{\theta}(1 - \exp(-\ell\theta t_p))\left(\frac{\ell^n}{p} + \mathfrak{c}\right) \\
&\leq \ell\mu(t_p + o(t_p^2))\left(\frac{\ell^n}{p} + \mathfrak{c}\right) \\
&\leq \ell\mu(1 + o(p))\left(\ell^n + p\mathfrak{c}\right) \\
&\leq \ell\mu\left(\ell^n + p\mathfrak{c} + o(p)\right),
\end{aligned} \tag{7.112}$$

where $\mathfrak{c} = \sup_{0 \leq t_p \leq p}\|\mathfrak{x}_1'(t_p)\|$. When $p \to 0^+$, it can be obtained that

$$\begin{aligned}
\|\mathfrak{X}'(p)\| &\leq \left\|\int_0^p \exp(\ell\Psi_{\mathfrak{x}}'(p - t_p))\ell\mathfrak{L}\|\mathfrak{f}(t_p)\|\mathrm{d}t_p\right\| \\
&\leq \left\|(\exp(\ell\Psi_{\mathfrak{x}}')p - I_{n+1})\left[\bar{\Psi}_{\mathfrak{x}}'^\dagger \; 0_{n+1}\right]\mathfrak{L} \cdot \ell\mu\left(\ell^n + p\mathfrak{c} + o(p)\right)\right\| \\
&\leq \left\|\left(\ell\Psi_{\mathfrak{x}}'p + o(p^2)\right)\left[\bar{\Psi}_{\mathfrak{x}}'^\dagger \; 0_{n+1}\right]\mathfrak{L} \cdot \ell\mu\left(\ell^n + p\mathfrak{c} + o(p)\right)\right\| \\
&\leq \ell^{n+2}p\|\mathfrak{L}\|\mu(1 + o(p)),
\end{aligned} \tag{7.113}$$

where

$$\Psi_{\mathfrak{x}}' \triangleq \begin{bmatrix} \bar{\Psi}_{\mathfrak{x}}' \\ 0_{n+1}^\mathsf{T} \end{bmatrix}, \tag{7.114}$$

$\bar{\Psi}_{\mathfrak{x}}'^\dagger$ is the pseudo-inverse of $\Psi_{\mathfrak{x}}'$. By combining (7.110) and (7.113), (7.107) is derived, i.e., $\|\mathfrak{X}'(p)\|/\|\mathfrak{X}(p)\| \ll 1$, $p \to 0^+$. Theorem 7.4 has been proved.

Remark 7.4 Theorem 7.4 indicates that the saturated observer has a good suppression performance for measurement noise. Especially when the noise has a pulse characteristic, the observation error of the AFTC saturation design is smaller, i.e., $\gamma'(t)$ is less than $\gamma(t)$. The fault detection logic (7.85) is less affected by the noise term $\ell^n \bar{\upsilon}$, and the small observation error also means that the corresponding tracking error is small. Therefore, the saturated structure is superior to the general structure.

7.4.2 Dead-Zone Redesign of Fault-Tolerant Control

7.4.2.1 Single Fault: Actuator

In this section, the FTC structure is redesigned from the perspective of duality by utilizing the dynamic dead-zone bound. The m dead-zone observers are redesigned as

$$
\begin{cases}
\dot{\hat{\xi}}''_j = \begin{bmatrix} \hat{\xi}''_{j,2} \\ \vdots \\ \hat{\xi}''_{j,r_j} \\ F^j_o(\hat{\xi}''_j) + \hat{\delta}''_j + G^j_o(\hat{\xi}''_j)u''_1 \end{bmatrix} + \begin{bmatrix} \frac{L_{j,1}}{\ell^{r_j-1}}\mathrm{dz}_{\sigma_j}(\varepsilon''_j) \\ \frac{L_{j,2}}{\ell^{r_j-2}}\mathrm{dz}_{\sigma_j}(\varepsilon''_j) \\ \vdots \\ L_{j,r_j}\mathrm{dz}_{\sigma_j}(\varepsilon''_j) \end{bmatrix}, \\
\dot{\hat{\delta}}''_j = \ell L_{j,r_j+1}\mathrm{dz}_{\sigma_j}(\varepsilon''_j), \\
\dot{\sigma}_j = -\ell\theta\sigma_j + \ell\mu\|\varepsilon''_j\|, \sigma_j \geq 0,
\end{cases} \tag{7.115}
$$

where σ_j represents a time-varying dead zone, θ, μ are two positive constants, $\varepsilon''_j = \ell^{r_j}(y_j - \hat{\xi}''_{j,1})$, $\mathrm{dz}_{\sigma_j}(\varepsilon''_j) = \varepsilon''_j - \mathrm{Sat}_{\sigma_j}(\varepsilon''_j)$, $\hat{\xi}''_j = [\hat{\xi}''_{j,1}, \hat{\xi}''_{j,2}, \cdots, \hat{\xi}''_{j,r_j}]^\mathsf{T}$ is the dead-zone observation, and $\hat{\delta}''_j$ the actuator fault estimation under the jth measurement. The overall observation vector is $\hat{\xi}'' = [\hat{\xi}''^\mathsf{T}_1, \hat{\xi}''^\mathsf{T}_2, \cdots, \hat{\xi}''^\mathsf{T}_m]$, and the corresponding estimation $\hat{X}''$ is

$$
\hat{X}'' = \mathrm{INV}(\Upsilon_{[m]})(\hat{\xi}'') \triangleq \mathrm{Sat}_{\Xi_x}\left(\Upsilon^{-1}_{[m]}\left([\mathrm{Sat}_{\Xi_1}\hat{\xi}''_1, \mathrm{Sat}_{\Xi_2}\hat{\xi}''_2, \cdots, \mathrm{Sat}_{\Xi_m}\hat{\xi}''_m]^\mathsf{T}\right)\right). \tag{7.116}
$$

The fusion estimation of the actuator fault is

$$
\hat{\delta}'' = \frac{\sum_{j\in[m]}\hat{\delta}''_j}{m}. \tag{7.117}
$$

Therefore, under the dead-zone strategy, an FTC framework is redesigned as follows

$$
u''_1 = -G^{-1}(\hat{X}'')\left[A_{0\sim n-1}\left(\hat{X}'' - x_r^{(0\sim n-1)}\right) + F(\hat{X}'') + \hat{\delta}'' - x_r^{(n)}\right]. \tag{7.118}
$$

Theorem 7.5 *For the nonlinear FAS (7.1) with only the actuator fault, considering the FTC dead-zone design (7.115)-(7.118), the state estimation error, actuator fault estimation error and tracking error are ultimately uniformly bounded, i.e., there exist positive constants t_1'', t_2'', M_1'', M_2'', M_3'', M_4'', M_5'', M_6'' such that*

$$
\begin{cases}
\|x^{(0\sim n-1)} - \hat{X}''\| \le \dfrac{M_1''}{\ell^2} + \ell^{n-1} M_2'' \bar{v}, \forall t > t_1'', \\[2mm]
\|\delta - \hat{\delta}''\| \le \dfrac{M_3''}{\ell} + \ell^n M_4'' \bar{v}, \forall t > t_1'', \\[2mm]
\|E\| \le \dfrac{M_5''}{\ell} + \ell^n M_6'' \bar{v}, \forall t > t_2''.
\end{cases}
\tag{7.119}
$$

Furthermore, if there is no measurement noise, i.e., $\bar{v} = 0$, then the aforementioned error can converge to a tiny neighborhood around the origin.

Proof For the measurement y_j, define the coordinate transformation

$$
\begin{aligned}
\zeta_j'' &= [\zeta_{j,1}'', \zeta_{j,2}'', \cdots, \zeta_{j,r_j}'', \zeta_{j,r_j+1}'']^{\mathsf{T}} \\
&= [\ell^{r_j}(\xi_{j,1} - \hat{\xi}_{j,1}''), \ell^{r_j-1}(\xi_{j,2} - \hat{\xi}_{j,2}''), \cdots, \delta - \hat{\delta}_j'']^{\mathsf{T}},
\end{aligned}
\tag{7.120}
$$

and the derivative of ζ_j'' is

$$
\begin{cases}
\dot{\zeta}_{j,k}'' = \ell\zeta_{j,k+1}'' - \ell L_{j,k}\zeta_{j,1}'' + \ell L_{j,k}(\mathfrak{q}_j + \mathfrak{Q}_j), k \in [r_j - 1], \\
\dot{\zeta}_{j,r_j}'' = \ell\zeta_{j,r_j+1}'' - \ell L_{j,r_j}\zeta_{j,1}'' + \ell L_{j,r_j}(\mathfrak{q}_j + \mathfrak{Q}_j) + \varpi_j'', \\
\dot{\zeta}_{j,r_j+1}'' = \dot{\delta} - \ell L_{j,r_j+1}\zeta_{j,1}'' + \ell L_{j,r_j+1}(\mathfrak{q}_j + \mathfrak{Q}_j), \\
\dot{\sigma}_j = -\ell\theta\sigma_j + \ell\mu\|\zeta_{j,1}''\| + \ell^{r_j} v_j\|,
\end{cases}
\tag{7.121}
$$

where

$$
\mathfrak{q}_j = \zeta_{j,1}'' - \mathrm{dz}_{\sigma_j}(\zeta_{j,1}'') = \mathrm{Sat}_{\sigma_j}(\zeta_{j,1}''), \tag{7.122}
$$

$$
\mathfrak{Q}_j = \mathrm{dz}_{\sigma_j}(\zeta_{j,1}'') - \mathrm{dz}_\sigma(\zeta_{j,1}'' + \ell^{r_j} v_j), \tag{7.123}
$$

$$
\varpi_j'' = \ell[F_o^j(\xi_j) - F_o^j(\hat{\xi}'') + (G_o^j(\xi_j) - G_o^j(\hat{\xi}_j''))u_1'']. \tag{7.124}
$$

The dynamics of E is

$$
\begin{cases}
\dot{e}_i = e_{i+1}, i \in [n-1], \\
\dot{e}_n = \rho'' + A_{0\sim n-1}(X - \hat{X}'') - A_{0\sim n-1}E,
\end{cases}
\tag{7.125}
$$

where

$$
\rho'' = F(x^{0\sim n-1}) - F(\hat{X}'') + \delta - \hat{\delta}'' + [G(x^{(0\sim n-1)}) - G(\hat{X}'')]u_1''. \tag{7.126}
$$

Therefore, the overall error dynamics is

$$
\begin{cases}
\dot{\zeta}''_{j,k} = \ell\zeta''_{j,k+1} - \ell L_{j,k}\zeta''_{j,1} + \ell L_{j,k}(\mathfrak{q}_j + \mathfrak{Q}_j), \\
\dot{\zeta}''_{j,r_j} = \ell\zeta''_{j,r_j+1} - \ell L_{j,r_j}\zeta''_{j,1} + \ell L_{j,r_j}(\mathfrak{q}_j + \mathfrak{Q}_j) + \varpi''_j, \\
\dot{\zeta}''_{j,r_j+1} = \dot{\delta} - \ell L_{j,r_j+1}\zeta''_{j,1} + \ell L_{j,r_j+1}(\mathfrak{q}_j + \mathfrak{Q}_j), \\
\dot{\sigma}_j = -\ell\theta\sigma_j + \ell\mu\|\zeta''_{j,1}\| + \ell^{r_j}v_j\|, \\
\dot{e}_i = e_{i+1}, i \in [n-1], \\
\dot{e}_n = \rho'' + A_{0\sim n-1}\tilde{X}'' - A_{0\sim n-1}E,
\end{cases}
\tag{7.127}
$$

where $j \in [m], k \in [r_j - 1], \tilde{X}'' = X - \hat{X}''$. It follows that

$$
\begin{cases}
\|\tilde{X}''\| \le a''_1\|\zeta''\|, \|\mathfrak{q}_j\| \le \|\zeta''_{j,1}\|, \|\varpi''_j\| \le a''_2\|\zeta''\|, \|\rho''\| \le a''_2\|\zeta''\|, \\
\|\mathfrak{Q}_j\| = \|\mathrm{Sat}_{\sigma_j}(\zeta''_{j,1}) - \mathrm{Sat}_{\sigma_j}(\zeta''_{j,1} + \ell^{r_j}v_j) - \ell^{r_j}v_j\| \le 2\|\ell^{r_j}v_j\| \le 2\ell^n\bar{v},
\end{cases}
\tag{7.128}
$$

where a''_1, a''_2 are positive constants.

Consider a Lyapunov function

$$
V''_1 = \sum_{j\in[m]} \zeta''^{\mathsf{T}}_j \mathscr{P}_j\zeta''_j + \sum_{j\in[m]}\sigma_j + E^{\mathsf{T}}PE.
\tag{7.129}
$$

The Dini derivative of V''_1 is

$$
\begin{aligned}
\mathrm{D}^+V''_1 = &\sum_{j\in[m]}\left\{\ell\zeta''^{\mathsf{T}}_j(\mathscr{P}_j\Psi_j + \Psi^{\mathsf{T}}_j\mathscr{P}_j)\zeta''_j - 2\ell(\mathfrak{q}_j + \mathfrak{Q}_j)L_j\mathscr{P}_j\zeta''_j\right. \\
&\left. + 2\dot{\delta}B^{\mathsf{T}}_{r_j+1}\mathscr{P}_j\zeta''_j + 2\varpi''_j\mathfrak{B}^{\mathsf{T}}_{r_j}\mathscr{P}_j\zeta''_j\right\} \\
&+ \sum_{j\in[m]}\left\{-\ell\theta\sigma_j + \ell\mu\|\zeta''_{j,1}\| + \ell^{r_j}v_j\|\right\} \\
&+ E^{\mathsf{T}}(P\Phi_A + \Phi^{\mathsf{T}}_A P)E + 2\rho''B^{\mathsf{T}}_{r_j}PE + 2A_{0\sim n-1}\tilde{X}''B^{\mathsf{T}}_3PE \\
= &-\ell\kappa_2\|\zeta''\|^2 - \kappa_1\|E\|^2 + 2\rho''B^{\mathsf{T}}_{r_j}PE + 2A_{0\sim n-1}\tilde{X}''B^{\mathsf{T}}_3PE \\
&+ \sum_{j\in[m]}\left\{2\dot{\delta}B^{\mathsf{T}}_{r_j+1}\mathscr{P}_j\zeta''_j + 2\varpi''_j\mathfrak{B}^{\mathsf{T}}_{r_j}\mathscr{P}_j\zeta''_j - 2\ell(\mathfrak{q}_j + \mathfrak{Q}_j)L_j\mathscr{P}_j\zeta''_j\right\} \\
&+ \sum_{j\in[m]}\left\{-\ell\theta\sigma_j + \ell\mu\|\zeta''_{j,1}\| + \ell^{r_j}v_j\|\right\}.
\end{aligned}
\tag{7.130}
$$

Based on Assumption 7.1 and (7.128), there exist positive constants $a_3'', a_4'', a_5'', a_6''$ such that

$$
\begin{aligned}
D^+ V_1'' &< -\ell\kappa_2 \|\zeta''\|^2 - \kappa_1 \|E\|^2 + a_3'' \|\zeta''\| + a_4'' \|\zeta''\|^2 \\
&\quad + 2\ell a_5''(\|\zeta''\| + 2\ell^n \bar{v})\|\zeta''\| + a_6'' \|\zeta''\|\|E\| \\
&\quad - \sum_{j=1}^m \ell\theta\sigma_j + \sum_{j=1}^m \ell\mu\|\zeta''\| + \sum_{j=1}^m \ell^{r_j+1}\mu\bar{v} \\
&< -(\ell\kappa_2 - a_7'')\|\zeta''\|^2 - \kappa_1 \|E\|^2 + (\ell^{n+1}a_9'' + a_8'')\|\zeta''\| \\
&\quad + a_6'' \|\zeta''\|\|E\| - \ell\theta\sum_{j=1}^m \sigma_j + \ell^{n+1}m\mu\bar{v},
\end{aligned}
\tag{7.131}
$$

where

$$
a_7'' = a_4'' + 2\ell a_5'', \; a_8'' = a_3'' + \ell m\mu, \; a_9'' = 4a_5''\bar{v}.
\tag{7.132}
$$

Moreover, the Dini derivative of V_1'' satisfies

$$
\begin{aligned}
D^+ V_1'' &< -\left(\frac{\ell\kappa_2}{2} - a_7''\right)\|\zeta''\|^2 + \left(\ell^{n+1}a_9'' + a_8''\right)\|\zeta''\| \\
&\quad - \left(\kappa_1 - \frac{a_6''^2}{2\ell\kappa_2}\right)\|E\|^2 + \ell^{n+1}m\mu\bar{v}.
\end{aligned}
\tag{7.133}
$$

If ℓ satisfies

$$
\ell > \max\left\{1, \frac{2a_7''}{\kappa_2}, \frac{a_6''}{2\kappa_1\kappa_2}\right\},
\tag{7.134}
$$

then the error $[\zeta''^\mathsf{T}, E''^\mathsf{T}, \sigma_1, \sigma_2, \cdots, \sigma_m]^\mathsf{T}$ is ultimately uniformly bounded. When there is no noise in the system, if the parameters ℓ, κ_2, κ_1 are set appropriately, the error can converge to a tiny neighborhood around the origin. The remaining proof is similar to Lemma 7.3, and the error system satisfies (7.119). Theorem 7.5 has been proved.

Remark 7.5 Parallel to the FTC saturation design, the FTC dead-zone design (7.118) exhibits similar performance to the FTC general design (7.15). The main reason is that the deterministic structure cannot completely eliminate the influence of measurement noise. However, in a physical sense, since the dead-zone observer will place error caused by noise within the dead zone, this novel structure has certain noise suppression performance. The dead-zone design has a significant inhibitory effect on persistent noise [15].

7.4.2.2 Multiple Faults: Actuator and Sensor

Define

$$
\gamma''(t) = \sum_{j \in [m]} \max \left\{ \lambda''_j \exp(-\alpha''_j \ell t), \frac{\Lambda''_j}{\ell^2} + \ell^{n-1} \beta''_j \bar{v} \right\}, \tag{7.135}
$$

where $\beta''_j, \alpha''_j, \Lambda''_j$ are positive constants. The observation error under actuator and sensor faults can be modeled as

$$
\xi - \hat{\xi}'' = \mathfrak{e}'' + \mathfrak{r}''_{[m]}, \tag{7.136}
$$

where $\mathfrak{e}''$ represents the error unrelated to sensor faults satisfying $\|\mathfrak{e}''\| \le \gamma''(t)$, and $\mathfrak{r}''_{[m]}$ represents the error related to sensor faults. For the $2d$-redundant observable system (7.1) with at most d sensor faults, a fault residual signal has been redesigned based on the sensor index set I, $|I| = m - d$:

$$
\mathfrak{R}''_I = \Upsilon_I \left(\mathrm{INV}(\Upsilon_I)(\hat{\xi}''_I) \right) - \hat{\xi}''_I. \tag{7.137}
$$

Theorem 7.6 *If*

$$
\|\mathfrak{R}''_I\| > \overline{\mathrm{Lip}}(\Upsilon_I \circ \mathrm{INV}(\Upsilon_I) - \mathscr{I})\gamma''(t), \tag{7.138}
$$

then there is a sensor fault in the index set I. Conversely, if for a $2d$-redundant observable system (7.1) with at most d sensor faults, (7.138) does not hold, the following state observer, actuator fault estimator, and sensor fault estimator are designed as

$$
\hat{X}''_I = \mathrm{INV}(\Upsilon_I)(\hat{\xi}''_I), \ \hat{\delta}''_I = \frac{\sum_{j \in I} \hat{\delta}''_j}{m - d}, \ \hat{s}''_{j,I} = y_j - H_j(\hat{X}''_I), j \in [m]. \tag{7.139}
$$

At this point, there exist positive constants t''_3, M''_{I1}, and M''_{I2} such that

$$
\|x^{(0 \sim n-1)} - \hat{X}''_I\| \le \frac{M''_{I1}}{\ell^2} + \ell^{n-1} M''_{I2} \bar{v}, \forall t > t''_3, \tag{7.140}
$$

where $\hat{X}''_I, \hat{\delta}''_I, \hat{s}''_{j,I}$ are the state observation, actuator fault estimation and sensor fault estimation under the sensor set I, respectively.

Proof The proof process is similar to Lemma 7.5.

If

$$
\|\mathfrak{R}''_{\iota(t)}\| > \overline{\mathrm{Lip}}(\Upsilon_{\iota(t)} \circ \mathrm{INV}(\Upsilon_{\iota(t)}) - \mathscr{I})\gamma'', \tag{7.141}
$$

then the dynamic sensor set $\iota(t)$ is updated to the next set $\iota(t^+)$. The AFTC dead-zone design is

$$
\begin{cases}
\hat{X}''_{\iota(t)} = \mathrm{INV}(\Upsilon_{\iota(t)})(\hat{\xi}''_{\iota(t)}), \hat{\delta}''_{\iota(t)} = \dfrac{\sum_{j \in \iota(t)} \hat{\delta}''_j}{m - d}, \hat{s}''_{j,\iota(t)} = y_j - H_j(\hat{X}''_{\iota(t)}), j \in [m], \\
u''_2 = -G^{-1}(\hat{X}''_{\iota(t)}) \left[A_{0 \sim n-1} \left(\hat{X}''_{\iota(t)} - x_r^{(0 \sim n-1)} \right) + F(\hat{X}''_{\iota(t)}) + \hat{\delta}''_{\iota(t)} - x_r^{(n)} \right],
\end{cases}
\tag{7.142}
$$

where $\hat{X}''_{\iota(t)}, \hat{\delta}''_{\iota(t)}, \hat{s}''_{j,\iota(t)}$ are the state observation, actuator fault estimation, and sensor fault estimation under the sensor set $\iota(t)$, respectively.

Theorem 7.7 *For the nonlinear FAS (7.1) with actuator and sensor faults, considering the AFTC dead-zone design (7.142), the state estimation error, fault estimation error, and tracking error are ultimately uniformly bounded, i.e., there exist positive constants* $t''_4, t''_5, M''_7, M''_8, M''_9, M''_{10}, M''_{11}, M''_{12}, M''_{13}, M''_{14}$ *such that*

$$
\begin{cases}
\| x^{(0 \sim n-1)} - \hat{X}''_{\iota(t)} \| \le \dfrac{M''_7}{\ell^2} + \ell^{n-1} M''_8 \bar{v}, \forall t > t''_4, \\
\| \delta - \hat{\delta}''_{\iota(t)} \| \le \dfrac{M''_9}{\ell} + \ell^n M''_{10} \bar{v}, \forall t > t''_4, \\
\| s_j - \hat{s}''_{j,\iota(t)} \| \le \dfrac{M''_{11}}{\ell^2} + \ell^{n-1} M''_{12} \bar{v}, j \in [m], \forall t > t''_4, \\
\| E \| \le \dfrac{M''_{13}}{\ell} + \ell^n M''_{14} \bar{v}, \forall t > t''_5.
\end{cases}
\tag{7.143}
$$

Furthermore, if there is no measurement noise, i.e., $\bar{v} = 0$, the aforementioned error can converge to a tiny neighborhood around the origin.

Proof When both the actuator and sensor fail simultaneously, a Lyapunov function is chosen as

$$
V''_2 = \sum_{j \in U} \zeta''^{\mathsf{T}}_j \mathscr{P}_j \zeta''_j + \sum_{j \in U} \sigma_j + E^{\mathsf{T}} P E.
\tag{7.144}
$$

The remaining stability proof is similar to Theorem 7.5.

7.4.2.3 Performance Comparison Analysis of Linear FASs

The comparative analysis of linear systems can to some extent reflect the advantages of dead-zone design. Since the core of both dead-zone design and general design is the ideal state observation, the DC gain from noise to estimation error can be used as a performance indicator for comparison.

Consider the following linear FAS model:

$$
\begin{cases}
x^{(n)} = \mathscr{A} x^{(0 \sim n-1)} + \mathscr{B} u + \delta, \\
y_1 = x + v_1,
\end{cases}
\tag{7.145}
$$

where $x, u, y_1 \in \mathbb{R}$ are the system state, the control input, and the measurement output, respectively. $\mathscr{A} = [\mathscr{A}_1, \mathscr{A}_2, \cdots, \mathscr{A}_n] \in \mathbb{R}^{1 \times n}$, $\mathscr{B} \in \mathbb{R}$ are known, v_1 is noise, and δ is assumed to be a constant fault.

The noise suppression performance mainly depends on the observation performance. The observer form in the general design, as shown in (7.93), and the derivative of $\mathfrak{X}$ is

$$\dot{\mathfrak{X}} = \ell \Psi_{\mathfrak{x}} \mathfrak{x} - \ell^{n+1} \mathfrak{L} v_1. \tag{7.146}$$

Correspondingly, the observer in the dead-zone design is

$$\begin{cases} \dot{\hat{X}}'' = \begin{bmatrix} \hat{x}_2'' \\ \vdots \\ \hat{x}_n'' \\ \mathscr{A}\hat{X}'' + \hat{\delta}'' + \mathscr{B}u \end{bmatrix} + \begin{bmatrix} \frac{\mathfrak{L}_1}{\ell^{n-1}} \mathrm{dz}_{\sigma_1}(\varepsilon'') \\ \frac{\mathfrak{L}_2}{\ell^{n-2}} \mathrm{dz}_{\sigma_1}(\varepsilon'') \\ \vdots \\ \mathfrak{L}_n \mathrm{dz}_{\sigma_1}(\varepsilon'') \end{bmatrix}, \\ \dot{\hat{\delta}}'' = \ell \mathfrak{L}_{n+1} \mathrm{dz}_{\sigma_1}(\varepsilon''), \\ \dot{\sigma}_1 = -\ell\theta\sigma_1 + \ell\mu \|\varepsilon''\|, \sigma_1 \geq 0, \end{cases} \tag{7.147}$$

where $\varepsilon'' = \ell^n(y_1 - \hat{x}_1'')$, and σ_1 is a dynamic dead-zone bound. Define the observation error

$$\begin{aligned} \mathfrak{X}'' &= [\mathfrak{x}_1'', \mathfrak{x}_2'', \cdots, \mathfrak{x}_n'', \mathfrak{x}_{n+1}'']^{\mathsf{T}} \\ &= [\ell^n(x_1 - \hat{x}_1''), \ell^{n-1}(x_2 - \hat{x}_2''), \cdots, \delta - \hat{\delta}'']^{\mathsf{T}}, \end{aligned} \tag{7.148}$$

and the derivative of $\mathfrak{X}''$ is

$$\begin{cases} \dot{\mathfrak{X}}'' = \ell \Psi_{\mathfrak{x}}' \mathfrak{X}'' - \ell \mathfrak{L} \mathrm{dz}_{\sigma_1}(\mathfrak{x}_1'' + \ell^n v_1), \\ \dot{\sigma}_1 = -\ell\theta\sigma_1 + \ell\mu \|\mathfrak{x}_1'' + \ell^n v_1\|, \end{cases} \tag{7.149}$$

where $\Psi_{\mathfrak{x}}'$ refers to (7.105).

Theorem 7.8 *If it is assumed that the measurement noise v_1 is a constant perturbation, and the matrix $\Psi_{\mathfrak{x}}$ is invertible, then the DC gain from noise to estimation error in (7.146) and (7.149) satisfies*

$$\frac{\mathrm{DC}_{\mathrm{dz}}}{\mathrm{DC}_0} = 1 - \frac{\mu}{\theta}, \forall \theta > \mu \geq 0, \tag{7.150}$$

where $\mathrm{DC}_{\mathrm{dz}}$, DC_0 are the DC gain in the dead-zone design and the general design, respectively.

Proof In (7.146), the equilibrium point of $\mathfrak{X}$ is

$$\mathcal{X}^{\circ} = \ell^{n} \Psi_{\mathfrak{x}}^{-1} \mathfrak{L} v_{1}. \tag{7.151}$$

In Eq. (7.149), the equilibrium point of σ_{1} is

$$\sigma_{1}^{\circ} = \frac{\mu}{\theta} \| C_{n+1} \mathcal{X}''^{\circ} + \ell^{n} v_{1} \| < \| C_{n+1} \mathcal{X}''^{\circ} + \ell^{n} v_{1} \|, \tag{7.152}$$

where $C_{n+1} = [1, 0, \cdots, 0] \in \mathbb{R}^{1 \times (n+1)}$, and $\mathcal{X}''^{\circ}$ is the equilibrium point of $\mathcal{X}''$. Then, the equilibrium point $\mathcal{X}''^{\circ}$ satisfies

$$\Psi_{\mathfrak{x}}' \mathcal{X}''^{\circ} - \mathfrak{L}(C_{n+1} \mathcal{X}''^{\circ} + \ell^{n} v_{1}) \left(1 - \frac{\mu}{\theta} \right) = 0. \tag{7.153}$$

According to the Woodbury matrix identity, it follows that

$$\begin{aligned} \mathcal{X}''^{\circ} &= \ell^{n} \left(1 - \frac{\mu}{\theta} \right) \left(\Psi_{\mathfrak{x}}' - \mathfrak{L} C_{n+1} \right)^{-1} \mathfrak{L} v_{1} \\ &= \ell^{n} \left(1 - \frac{\mu}{\theta} \right) \Psi_{\mathfrak{x}}^{-1} \mathfrak{L} v_{1}. \end{aligned} \tag{7.154}$$

Therefore, the DC gain from noise to estimation error in (7.146) and (7.149) satisfies

$$\frac{\mathrm{DC}_{dz}}{\mathrm{DC}_{0}} = \frac{\| \mathcal{X}''^{\circ} \|}{\| \mathcal{X}^{\circ} \|} = 1 - \frac{\mu}{\theta}. \tag{7.155}$$

Theorem 7.8 has been proved.

Remark 7.6 $\mu = 0$ indicates that the dead-zone strategy is not activated, and in this case, the observer degenerates to the general form. The smaller the DC gain, the better the noise suppression performance. The suppression performance of the dead-one design increases as μ increases, but a larger μ will lead to a larger system error. Therefore, the parameter adjustments of μ and μ/θ should be determined based on different task environments.

7.5 Simulation and Comparative Analysis

Consider a third-order FAS model:

$$\begin{cases} x^{(3)} = 3 - 3\ddot{x} - 2\dot{x} - \sin(\ddot{x} + 2\dot{x} - 2) + u + \delta, \\ y_{1} = x + s_{1} + v_{1}, \\ y_{2} = x + s_{2} + v_{2}, \\ y_{3} = \dot{x} + s_{3} + v_{3}, \\ y_{4} = 2\dot{x} + 4x - 2 + s_{4} + v_{4}, \end{cases} \tag{7.156}$$

where

$$\delta = -\rho u + u_f. \tag{7.157}$$

The time-varying fault models are

$$\rho(t) = \begin{cases} 0, t < 10\text{s}, \\ 0.2, t \geq 10\text{s}, \end{cases} \quad u_f(t) = \begin{cases} 0, t < 10\text{s}, \\ \sin(t - 10), t \geq 10\text{s}, \end{cases} \tag{7.158}$$

and

$$s_1 = s_3 = s_4 = 0, s_2 = \begin{cases} 0, t < 20\text{s}, \\ (t - 20) \sin t, t \geq 20\text{s}. \end{cases} \tag{7.159}$$

The observable spaces corresponding to the four measurements are

$$\begin{cases} \xi_1 = \Upsilon_1(x^{(0\sim2)}) = [x, \dot{x}, \ddot{x}]^\mathsf{T}, \\ \xi_2 = \Upsilon_2(x^{(0\sim2)}) = [x, \dot{x}, \ddot{x}]^\mathsf{T}, \\ \xi_3 = \Upsilon_3(x^{(0\sim2)}) = [\dot{x}, \ddot{x}]^\mathsf{T}, \\ \xi_4 = \Upsilon_4(x^{(0\sim2)}) = [2\dot{x} + 4x - 2, 2\ddot{x} + 4\dot{x}]^\mathsf{T}. \end{cases} \tag{7.160}$$

The relative orders of the four measurements are $r_1 = 3, r_2 = 3, r_3 = 2$, and $r_4 = 2$. The FAS (7.156) is 2-redundant observable. The specific forms of the four mappings Υ_i^{-1} can be referred to Table 6.1.

In the upcoming AFTC simulation experiment, the initial values are set as zero, the reference signal is $x_r = \sin t$, $x^{(0\sim2)} \in \Omega_x \triangleq \{\varsigma \mid \varsigma \in \mathbb{R}^3 : \|\varsigma\|_\infty \leq 10\}$, and $\xi \in \Omega_\xi \triangleq \{\varsigma \mid \varsigma \in \mathbb{R}^{10} : \|\varsigma\|_\infty \leq 20\}$. Some parameters are set as $\ell = 20, A^{0\sim2} = [2, 4, 3], L_1 = L_2 = [4, 7, 6, 2], L_3 = L_4 = [2, 4, 3]$. Since it is impossible to obtain an absolutely ideal fault detection threshold, the simplified fault detection threshold is chosen as 2 through Monte Carlo simulation and some basic model information.

7.5.1 An Experiment Under the AFTC General Design

When $v_j, j \in [4]$, are uniformly distributed on $[-0.001, 0.001]$, Figs. 7.1 and 7.2 present the corresponding signal trajectories of the tracking control experiments under the general framework. Clearly, the AFTC general design can effectively handle actuator and sensor faults, which demonstrates the rationality of Lemma 7.6. However, the observation of high-order states is prone to be affected by measurement noise. The actuator fault corresponds to the highest-order position, so the oscillation phenomenon of actuator fault estimation is more obvious. As shown in Fig. 7.2b, the fused state observation can provide effective sensor fault estimation.

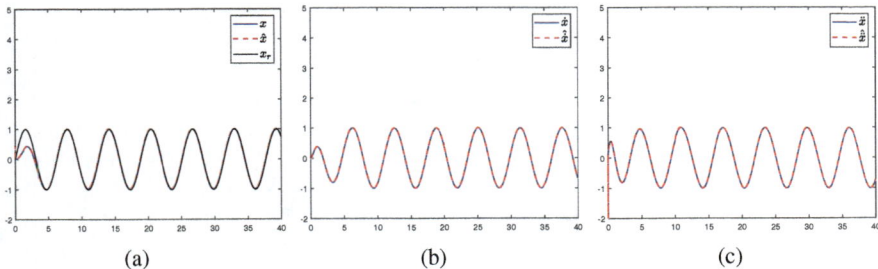

Fig. 7.1 The signal trajectories under the AFTC general design

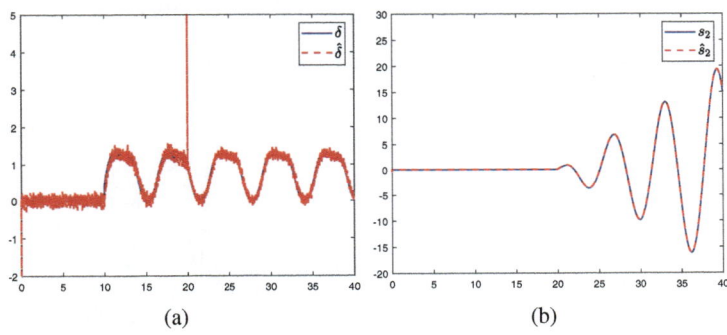

Fig. 7.2 The fault signals under the AFTC general design

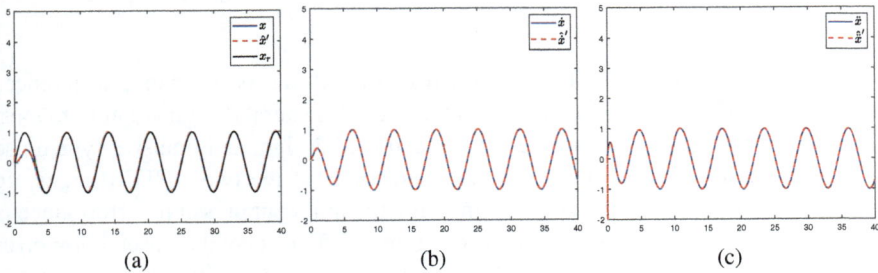

Fig. 7.3 The signal trajectories under the AFTC saturation design

7.5.2 An Experiment Under the AFTC Saturation Design

Under the same uniformly distributed noise, the parameter settings for the AFTC saturation design are $\theta = 0.1$ and $\mu = 0.9$. Figures 7.3 and 7.4 present the corresponding signal trajectories of the tracking control experiments under the AFTC saturation framework. It is evident that the AFTC saturation design can also effectively tolerate the two faults and provide effective actuator fault estimation and sensor fault estimation.

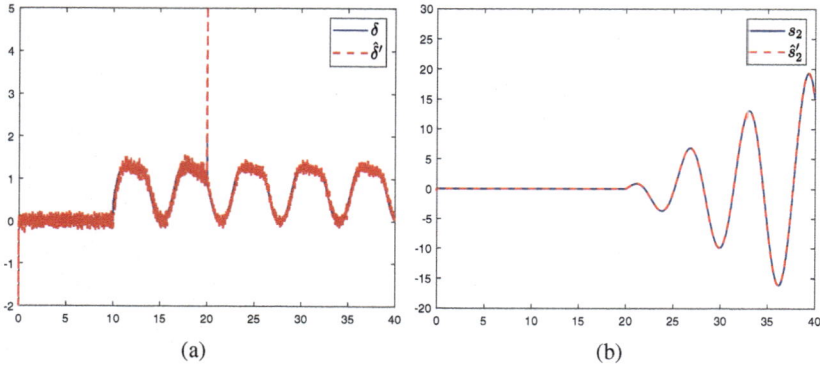

(a)   (b)

Fig. 7.4 The fault signals under the AFTC saturation design

Fig. 7.5 The observation error under impulsive-type noise based on general/saturation design

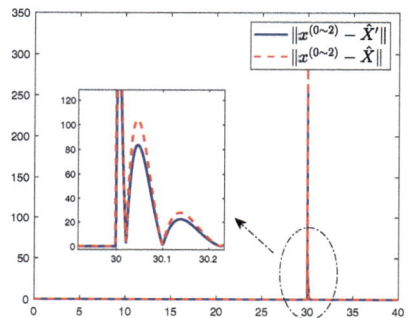

Under a uniformly distributed noise background, the performance differences between the AFTC saturation design and the AFTC general design are not obvious. To further highlight the advantages of the saturated design, an impulsive-type noise is injected at $t = 30$ s. Since both the foundations of the two AFTC designs are ideal state observations, the key to comparing the noise suppression performance is the analysis of the state observation error. Figure 7.5 presents the observation error under the two AFTC designs. At the moment of noise, the saturation structure is less sensitive, so the AFTC saturated design is more suitable for systems with noise, especially for nonlinear systems with impulsive-type noise.

7.5.3 An Experiment Under the AFTC Dead-Zone Design

At this point, v_j is still set to be uniformly distributed noise within the range of $[-0.001, 0.001]$. The parameter settings for the dead-one design are $\theta = 10$ and $\mu = 7$. From Figs. 7.6 and 7.7, it can be seen that the AFTC dead-zone design can also effectively compensate actuator and sensor faults. By comparing the experimental results of the general design with those of the dead-zone design, it

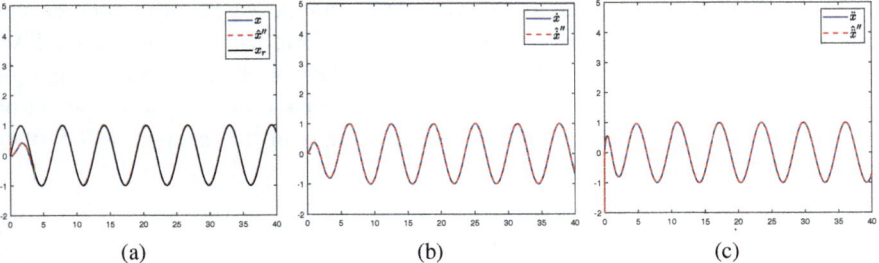

Fig. 7.6 The signal trajectories under the AFTC dead-zone design

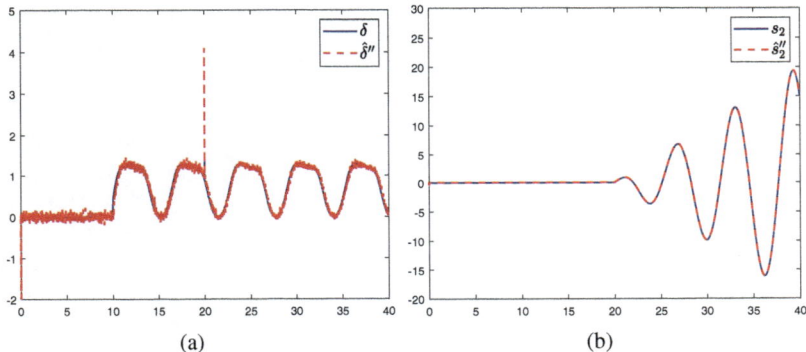

Fig. 7.7 The fault signals under the AFTC dead-zone design

is found that the oscillation phenomenon in Fig. 7.7a is weaker than that in Fig. 7.7b. More significantly, when the sensor fails, the jump phenomenon in the actuator fault estimation under the dead-zone design is weaker. Therefore, the AFTC dead-zone design structure has a better suppression effect on general noise.

Furthermore, numerous simulation experiments have shown that the AFTC saturation design offers better suppression performance for impulsive-type noise, while the AFTC dead-zone design provides better suppression performance for continuous-type noise.

7.6 Notes and References

This chapter establishes three AFTC frameworks based on the FAS approaches. From the perspective of the FAS theory, all three AFTC structures can maintain the parameterized characteristics of the controller, and the controller design process is universal and standardized. From the perspective of FTC theory, the AFTC structure can not only quickly compensate actuator faults, but also handle sudden and severe sensor faults. From the perspective of noise suppression, the AFTC saturation design can better suppress impulsive-type noise, and the AFTC dead-zone design can better suppress continuous-type noise. The parameterized structure in the FAS theory can provide a decoupled, global, and standard nonlinear controller design process. However, the requirement for accurate models limits the development of FAS controllers. The three integrated AFTC designs in this chapter will further promote the emerging FAS theory.

Chapter 8
FTC for Input-Nonaffine FASs

In this chapter, the input-nonaffine fully actuated system (FAS) controller structure is established for trajectory tracking and fault tolerance. From an unknown input-nonaffine FAS model, an observer-based control input dynamics is yielded, which ensures the low-power characteristics of the FAS controller. Both the observation error and tracking error eventually converge eventually, and the error bound can be adjusted to a tiny neighborhood around the origin. Moreover, for unknown input-nonaffine FASs with multiplicative actuator and sensor faults, a robust fault-tolerant control (FTC) strategy for stabilization tasks is proposed to realize the ultimately uniformly bounded stability. The main results are proved theoretically and illustrated experimentally.

8.1 Introduction to Input-Nonaffine FAS Models

The existing FAS theory mainly focuses on input-affine systems, while input-nonaffine systems represent a more general system structure. Professor Duan has initially proposed some ideas for the input-nonaffine FAS theory [30], but did not provide feasible controller design methods. The research on input-nonaffine FASs has almost come to a standstill. In the first-order state-space theory, controller design for input-nonaffine systems has made some progress, such as active disturbance rejection control [82], backstepping control [99], and neural network control [48]. However, these controller designs for input-nonaffine systems inevitably retain the limitations of first-order nonlinear state-space models.

The differential diffeomorphism in input-nonaffine FASs heavily relies on model information, but unmodeled dynamics, disturbances, and faults, among other factors, may make the precise system model impossible to be given. Moreover, the solution of the inverse function in input-nonaffine systems is much more difficult

© The Author(s) 2026
D. Zhou, M. Cai, *Fault-Tolerant Control for Fully Actuated Systems*,
https://doi.org/10.1007/978-981-95-0691-0_8

than the solution of the inverse matrix in input-affine systems. Therefore, the research on unknown input-nonaffine FASs is an urgent and challenging task.

This chapter preliminarily explores the structure of input-nonaffine FASs and proposes tracking controllers and robust fault-tolerant stabilization controllers for unknown input-nonaffine FASs. Starting from the unknown system model, a dynamic equation of the control input signal is given based on a state observer. Further, for the case of actuator and sensor faults in the unknown input-nonaffine FAS model, a robust fault-tolerant stabilization controller is proposed, ensuring the ultimate uniform bounded stability, and the error range can be adjusted to a tiny neighborhood of the origin.

8.2   Problem Formulation and Mathematical Modeling

Consider the following class of unknown input-nonaffine FASs:

$$
\begin{cases}
x^{(n)} = F(x^{(0 \sim n-1)}, u, \xi, d, t), \\
\dot{\xi} = F_0(x^{(0 \sim n-1)}, \xi, d, t), \\
y = x,
\end{cases}
\tag{8.1}
$$

where $x, u, y, d \in \mathbb{R}, \xi \in \mathbb{R}^m$ are the system state, the control input, the measurement output, and the unmodeled dynamics, respectively. $F(\cdot)$ and $F_0(\cdot)$ are unknown continuous differentiable functions, and the sign of $\frac{\partial F}{\partial u}$ is known. Unlike the uncertain input-affine FAS model, the input-nonaffine model (8.1) has stronger unknown characteristics.

Definition 8.1 (Input-Nonaffine FASs [28]) If $w \triangleq F(x^{(0 \sim n-1)}, u, \xi, d, t)$ is a differential diffeomorphism from u to w, then the nonlinear input-nonaffine system (8.1) is fully actuated, i.e., the input-nonaffine FASs satisfy

$$
\left\| \frac{\partial F}{\partial u}(x^{(0 \sim n-1)}, u, \xi, d, t) \right\| > 0, \forall t \geq 0.
\tag{8.2}
$$

If $\frac{\partial F}{\partial u}$ is a function independent of u, then the input-nonaffine system will degenerate into an input-affine system. Therefore, system (8.1) is a more general FAS model.

Assumption 8.1 The process disturbance satisfies $\sup_{t \geq 0} \|[d, \dot{d}]\| < \infty$.

Assumption 8.2 There exist two differentiable functions $\rho_1(x^{(0 \sim n-1)}, u, \xi, d)$ and $\rho_2(x^{(0 \sim n-1)}, \xi, d)$ such that for $\forall t \geq 0$, it holds that

$$
\begin{cases}
\max \left\{ \|F(x^{(0 \sim n-1)}, u, \xi, d, t)\|, \|\nabla F(x^{(0 \sim n-1)}, u, \xi, d, t)\| \right\} \leq \rho_1(x^{(0 \sim n-1)}, u, \xi, d), \\
\max \left\{ \|F_0(x^{(0 \sim n-1)}, \xi, d, t)\|, \|\nabla F_0(x^{(0 \sim n-1)}, \xi, d, t)\| \right\} \leq \rho_2(x^{(0 \sim n-1)}, \xi, d).
\end{cases}
\tag{8.3}
$$

Assumption 8.3 There exists a positive definite function $V_0(\xi)$ such that for $\forall \xi$: $\|\xi\| \geq \chi(\|(x^{(0\sim n-1)}, d)\|)$, it holds that

$$\frac{\partial V_0}{\partial \xi}(\xi) F_0(x^{(0\sim n-1)}, \xi, d, t) \leq 0, \tag{8.4}$$

where $\chi(\cdot)$ is a well-known $\mathcal{K}_\infty$ function.

Remark 8.1 Assumptions 8.1–8.3 are common requirements in nonlinear system observation and control [82, 83, 95]. In deterministic system models, many practical systems, e.g., robotic arm models, permanent magnet synchronous motor models, etc., all satisfy Assumptions 8.1–8.2. Generally, the Lipschitz condition is stricter than Assumption 8.2.

If the FAS model (8.1) is completely known, and the unmodeled dynamics and process disturbances are also known, then the input-nonaffine FAS controller [30]

$$\begin{cases} w = -A_{0\sim n-1} x^{(0\sim n-1)} + u_a \\ u = F^{-1}(x^{(0\sim n-1)}, \xi, d, w, t) \end{cases} \tag{8.5}$$

will give a linear system

$$\dot{x}^{(0\sim n-1)} = \Phi_A x^{(0\sim n-1)} + B_n u_a, \tag{8.6}$$

where $F^{-1}(\cdot)$ represents the inverse function of $F(\cdot)$, and u_a is an auxiliary signal used for trajectory tracking and other tasks,

$$\Phi_A = \begin{bmatrix} 0 & 1 & 0 & \cdots & 0 \\ 0 & 0 & 1 & \cdots & 0 \\ \vdots & & & \ddots & \vdots \\ 0 & 0 & 0 & \cdots & 1 \\ -A_0 & -A_1 & -A_2 & \cdots & -A_{n-1} \end{bmatrix}, \quad B_n = \begin{bmatrix} 0 \\ 0 \\ \vdots \\ 0 \\ 1 \end{bmatrix}. \tag{8.7}$$

This parametric structure enables the nonlinear system (8.1) to achieve global asymptotic stability. Moreover, the FAS model (8.1) can be transformed into

$$\begin{cases} \dot{x}^{(0\sim n-1)} = \Phi_0 x^{(0\sim n-1)} + B_n w, \\ y = x, \end{cases} \tag{8.8}$$

where

$$\Phi_0 = \begin{bmatrix} 0 & 1 & 0 & \cdots & 0 \\ 0 & 0 & 1 & \cdots & 0 \\ \vdots & & \ddots & & \vdots \\ 0 & 0 & 0 & \cdots & 1 \\ 0 & 0 & 0 & \cdots & 0 \end{bmatrix}. \tag{8.9}$$

In the transformed system model, w can be regarded as a virtual controller that is influenced by the actual controller u. The core issue of the input-nonaffine FAS theory is to solve the controller dynamics under the relationship between w and u, i.e., (8.5). However, in complex real-world objects, the unknown model makes it difficult to give the ideal controller (8.5). In input-nonaffine systems, due to the existence of uncertainties, disturbances, and faults, the inverse function $F^{-1}(\cdot)$ is also not easy to obtain. Therefore, it is necessary to study the controller design problem for unknown input-nonaffine FASs.

8.3 Tracking Control Design

The control objective of this section is trajectory tracking, i.e., to let x follow a smooth reference signal x_r that satisfies $\sup_{t \geq 0} \| x_r^{(0 \sim n+1)} \| < \infty$. At this point, the target trajectory of the reference signal x_r can be designed as

$$
\begin{cases}
\dot{x}_\star^{(0 \sim n-1)} = \Phi_0 x_\star^{(0 \sim n-1)} + B_n x_r^{(n)} - B_n A_{0 \sim n-1}[x_\star^{(0 \sim n-1)} - x_r^{(0 \sim n-1)}], \\
y_\star = x_\star,
\end{cases}
$$

(8.10)

where $x_\star^{(0 \sim n-1)} \in \mathbb{R}^n$, $y_\star \in \mathbb{R}$. When Φ_A is a Hurwitz matrix, it is easy to prove that $x_\star^{(0 \sim n-1)}$ asymptotically converges to $x_r^{(0 \sim n-1)}$. Naturally, the smooth signal $x_\star$ can become the new reference trajectory. By comparing (8.8) and (8.10), the ideal $w_\star$ is determined to satisfy

$$
w_\star = -A_{0 \sim n-1}(x^{(0 \sim n-1)} - x_r^{(0 \sim n-1)}) + x_r^{(n)}.
$$

(8.11)

According to input-nonaffine FAS controller (8.5), the auxiliary signal is designed as

$$
u_a = A_{0 \sim n-1} x_r^{(0 \sim n-1)} + x_r^{(n)}.
$$

(8.12)

This indicates that once the designed controller can ensure (8.11), the trajectory tracking task is accomplished.

However, the precise model (8.8) and the system state $x^{(0 \sim n-1)}$ are both unknown. Therefore, an observer approach is needed to estimate the system state and the unknown dynamics. Let $X = [x_1, x_2, \cdots, x_n]^\mathsf{T} \triangleq x^{(0 \sim n-1)}$, and an observer is designed as [43]

$$
\begin{cases}
\dot{\hat{x}}_i = \hat{x}_{i+1} + \kappa^{n-i} h_i \left(\dfrac{y - \hat{x}_1}{\kappa^n} \right), i \in [n-1], \\
\dot{\hat{x}}_n = \hat{w} + h_n \left(\dfrac{y - \hat{x}_1}{\kappa^n} \right), \\
\dot{\hat{w}} = \kappa^{-1} h_{n+1} \left(\dfrac{y - \hat{x}_1}{\kappa^n} \right),
\end{cases}
$$

(8.13)

where $\hat{x}_i$ represents the state observation of x_i, and $\hat{w}$ is the estimation of w. κ is a constant within the interval $(0, 1)$, and $h_i(\cdot), i = 1, 2, \cdots, n + 1$ are Lipschitz nonlinear functions that satisfies Assumption 8.4. The observer in Lemma 8.1 can be used to assist in the design of $h_i(\cdot)$.

Remark 8.2 Generally, the observer design involves both the control input and the measurement output. And w is a nonaffine function related to the control input u. The subsequent controller will ensure that $\hat{w}$ satisfies a specific trajectory

$$\hat{w} = -A_{0 \sim n-1} \hat{X} + u_a, \tag{8.14}$$

so $\hat{w}$ includes control input terms. Due to the structural characteristics of the input-nonaffine system (8.1), the input-output form cannot be explicitly expressed mathematically in the observer. Moreover, the unknown nature of w further makes it difficult for the observer to present a standard input-output form. This chapter can only indirectly reflect the input information in the form presented by (8.13), and similar ideas have also been successfully explored in some cases [66, 82, 98].

Assumption 8.4 There exist positive constants $a_1, a_2, a_3, a_4, \beta_1$ and two positive definite functions $V_1(\mathfrak{z}), W_1(\mathfrak{z}), \forall \mathfrak{z} = [\mathfrak{z}_1, \mathfrak{z}_2, \cdots, \mathfrak{z}_{n+1}]^\mathsf{T} \in \mathbb{R}^{n+1}$ such that

1. $a_1 \|\mathfrak{z}\|^2 \leq V_1(\mathfrak{z}) \leq a_2 \|\mathfrak{z}\|^2, a_3 \|\mathfrak{z}\|^2 \leq W_1(\mathfrak{z}) \leq a_4 \|\mathfrak{z}\|^2$.
2. $\sum\limits_{i=1}^{n} (\mathfrak{z}_{i+1} - h_i(\mathfrak{z}_1)) \dfrac{\partial V_1}{\partial \mathfrak{z}_i}(\mathfrak{z}) - h_{n+1}(\mathfrak{z}_1) \dfrac{\partial V_1}{\partial \mathfrak{z}_{n+1}}(\mathfrak{z}) \leq -W_1(\mathfrak{z})$.
3. $\left\| \dfrac{\partial V_1}{\partial \mathfrak{z}_{n+1}}(\mathfrak{z}) \right\| \leq \beta_1 \|\mathfrak{z}\|$.

Remark 8.3 Assumption 8.4 is a sufficient condition for the convergence of the nonlinear observer [43, 45]. Additionally, Corollary 8.1 provides a common structure meeting Assumption 8.4.

Based on (8.11), the controller dynamics is designed as

$$\dot{u} = -\ell \mathrm{sign} \left(\frac{\partial F}{\partial u} \right) \left[\hat{w} + A_{0 \sim n-1}(\hat{X} - x_r^{(0 \sim n-1)}) - x_r^{(n)} \right], \tag{8.15}$$

where ℓ is a constant greater than 1. Due to the limited energy of the actual actuator, the saturation function is utilized to redesign (8.15) as

$$\dot{u} = -\ell \mathrm{sign} \left(\frac{\partial F}{\partial u} \right) \left[\mathrm{Sat}_{M_n}(\hat{w}) - x_r^{(n)} + \sum_{i=0}^{n-1} A_i \mathrm{Sat}_{M_i}(\hat{x}_{i+1} - x_r^{(i)}) \right], \tag{8.16}$$

where $M_i, i = 0, 1, \cdots, n$ are designed constants. Due to the diverse bounds of the actuators, an analytical expression for M_i cannot be provided. The conservative value of M_i can be determined through practical mechanism models and computer simulation. To make the saturation controller more smooth, the following odd function is selected as [35]

$$\psi_\kappa(r) = \begin{cases} r, \ 0 \leq r \leq 1, \\ r + \dfrac{r-1}{\kappa^\kappa} - \dfrac{r^2-1}{2\kappa}, \ 1 \leq r \leq 1+\kappa, \\ 1 + \dfrac{\kappa}{2}, \ r \geq 1+\kappa, \end{cases} \tag{8.17}$$

This odd function satisfies

$$\left\| \frac{\mathrm{d}\psi_\kappa}{\mathrm{d}r}(r) \right\| \leq 1, \ \|\psi_\kappa(r) - \mathrm{Sat}_1(r)\| \leq \frac{r}{2}, \forall r \in \mathbb{R}. \tag{8.18}$$

Therefore, the controller dynamics (8.16) can be rewritten as

$$\dot{u} = -\ell \mathrm{sign}\left(\frac{\partial F}{\partial u}\right)\left[M_n\psi_\kappa\left(\frac{\hat{w}}{M_n}\right) - x_r^{(n)} + \sum_{i=0}^{n-1} A_i M_i \psi_\kappa\left(\frac{\hat{x}_{i+1} - x_r^{(i)}}{M_i}\right)\right]. \tag{8.19}$$

According to the FAS theory, if the parameter matrix Φ_A is a Hurwitz matrix, then there exists a positive definite solution $P \in \mathbb{R}^{n \times n}$ to the following Lyapunov equation

$$P\Phi_A + \Phi_A^\mathsf{T} P = -I_n. \tag{8.20}$$

Define

$$s = w + A_{0\sim n-1}(x^{(0\sim n-1)} - x_r^{(0\sim n-1)}) - x_r^{(n)}, \ V_2(s) = s^2, \tag{8.21}$$

$$\eta = [\eta_1, \eta_2, \cdots, \eta_n]^\mathsf{T} = x^{(0\sim n-1)} - x_r^{(0\sim n-1)}, \ V_3(\eta) = \eta^\mathsf{T} P\eta, \tag{8.22}$$

$$\varpi_1 = \sup_{[s,\eta^\mathsf{T}]^\mathsf{T} \in \mathbb{R}^{n+1}, \|[s,\eta^\mathsf{T}]^\mathsf{T}\| \leq \|[s(0),\eta(0)^\mathsf{T}]^\mathsf{T}\|} (V_2(s) + V_3(\eta)) + 1, \tag{8.23}$$

and two compact sets

$$\Omega_0 = \{[s, \eta^\mathsf{T}]^\mathsf{T} \mid [s, \eta^\mathsf{T}]^\mathsf{T} \in \mathbb{R}^{n+1}, V_2(s) + V_3(\eta) \leq \varpi_1\}, \tag{8.24}$$

$$\Omega_1 = \{[s, \eta^\mathsf{T}]^\mathsf{T} \mid [s, \eta^\mathsf{T}]^\mathsf{T} \in \mathbb{R}^{n+1}, V_2(s) + V_3(\eta) \leq \varpi_1 + 1\}. \tag{8.25}$$

Lemma 8.1 *For the input-nonaffine FAS model (8.1) under the observer (8.13) and the controller dynamics (8.19), if Assumptions 8.1–8.4 hold, then there exists $\ell^* > 1$ such that $\forall \ell > \ell^*, t \geq 0$, it holds that $[s(t), \eta(t)^\mathsf{T}]^\mathsf{T} \in \Omega_1$.*

Proof Based on $V_2(s(0)) + V_3(\eta(0)) < \varpi_1$ and the continuity of $s(t), \eta(t)$, it is obtained that

$$[s(0), \eta(0)^\mathsf{T}]^\mathsf{T} \in \Omega_0 - \partial\Omega_0, \tag{8.26}$$

where $\partial\Omega_0$ represents the boundary of the set Ω_0. Then, there exists $t_0 > 0$ such that for all $t \in [0, t_0]$, it holds that $([s(t), \eta(t)^\mathsf{T}]^\mathsf{T} \in \Omega_0 - \partial\Omega_0$. The remaining proof process is by contradiction. Assuming that Lemma 8.1 does not hold, i.e., for any $\check{\ell} > 1$, there exist $\ell > \check{\ell}$ and $t \geq 0$ such that

$$[s(t), \eta(t)^\mathsf{T}]^\mathsf{T} \in \mathbb{R}^{n+1} - \Omega_1. \tag{8.27}$$

$[s(t), \eta(t)^\mathsf{T}]^\mathsf{T}$ is a continuous function dependent on t, and for all $t \in [0, t_0]$, it holds that $[s(t), \eta(t)^\mathsf{T}]^\mathsf{T} \in \Omega_0 - \partial\Omega_0$, thus there exists $t_2 > t_1 > t_0$ such that

$$\begin{cases} [s(t_1), \eta(t_1)^\mathsf{T}]^\mathsf{T} \in \partial\Omega_0, \\ [s(t_2), \eta(t_2)^\mathsf{T}]^\mathsf{T} \in \Omega_1 - \Omega_0, \\ V_2(s(t_2)) + V_3(\eta(t_2)) > \varpi_1 + 0.5, \\ [s(t), \eta(t)^\mathsf{T}]^\mathsf{T} \in \Omega_1 - \Omega_0^\circ, \forall t \in [t_1, t_2], \\ [s(t), \eta(t)^\mathsf{T}]^\mathsf{T} \in \Omega_1, \forall t \in [0, t_2], \end{cases} \tag{8.28}$$

where Ω_0° represents the interior of the set Ω_0.

Define the observation error

$$\begin{aligned} e &= [e_1, e_2, \cdots, e_{n+1}]^\mathsf{T} \\ &= \left[\frac{x_1 - \hat{x}_1}{\kappa^n}, \frac{x_2 - \hat{x}_2}{\kappa^{n-1}}, \cdots, \frac{x_n - \hat{x}_n}{\kappa}, w - \hat{w} \right]^\mathsf{T}, \end{aligned} \tag{8.29}$$

and the derivative of e is

$$\begin{cases} \kappa\dot{e}_i = e_{i+1} - h_n(e_1), i \in [n], \\ \kappa\dot{e}_{n+1} = \kappa\hbar - h_{n+1}(e_1), \end{cases} \tag{8.30}$$

where $\hbar = \dot{w}$. During the time interval $[0, t_2]$, $\hbar$ can be calculated as

$$\hbar = \sum_{i=1}^{n-1} x_{i+1} \frac{\partial F}{\partial x_i} + \left(s - A_{0\sim n-1}\eta + x_r^{(n)} \right) \frac{\partial F}{\partial x_n}$$

$$+ \dot{u}\frac{\partial F}{\partial u} + \left(\frac{\partial F}{\partial \xi} \right)^\mathsf{T} F_0 + \dot{d}\frac{\partial F}{\partial d} + \frac{\partial F}{\partial t}. \tag{8.31}$$

From (8.19), it follows that

$$\dot{u} = \ell\operatorname{sign}\left(\frac{\partial F}{\partial u} \right) \left\{ x_r^{(n)} + H_n(\hat{w}) + \sum_{i=0}^{n-1} A_i H_i \left(\hat{x}_{i+1} - x_r^{(i)} \right) \right.$$

$$-\left[s - A_{0 \sim n-1}\eta + x_r^{(n)} - e_{n+1} + \sum_{i=0}^{n-1} A_i\left(\hat{x}_{i+1} - x_r^{(i)}\right)\right]\Bigg\}$$

$$= \ell\,\mathrm{sign}\left(\frac{\partial F}{\partial u}\right)\left[H_n(\hat{w}) + \sum_{i=0}^{n-1} A_i H_i\left(\hat{x}_{i+1} - x_r^{(i)}\right)\right.$$

$$\left. - \left(s - \sum_{i=0}^{n-1}\kappa^{n-i}A_i e_{i+1} - e_{n+1}\right)\right], \tag{8.32}$$

where

$$H_i(\hat{x}_{i+1} - x_r^{(i)}) = \hat{x}_{i+1} - x_r^{(i)} - M_i\psi_\kappa\left(\frac{\hat{x}_{i+1} - x_r^{(i)}}{M_i}\right), i = 0, 1, \cdots, n-1,$$

$$H_n(\hat{w}) = \hat{w} - M_n\psi_\kappa\left(\frac{\hat{w}}{M_n}\right). \tag{8.33}$$

The following is a detailed discussion on the odd function $H_n(\hat{w})$:

Case 1: $0 \le \hat{w} \le M_n$. It turns out that

$$\left\|H_n(\hat{w})\right\| = \left\|\hat{w} - M_n \cdot \frac{\hat{w}}{M_n}\right\| = 0. \tag{8.34}$$

Case 2: $M_n \le \hat{w} \le (1+\kappa)M_n$. It turns out that

$$\left\|H_n(\hat{w})\right\| = \left|\hat{w} - M_n\left(\frac{\hat{w}}{M_n} + \frac{\frac{\hat{w}}{M_n} - 1}{\kappa} - \frac{(\frac{\hat{w}}{M_n})^2 - 1}{2\kappa}\right)\right\|$$

$$= M_n\frac{(\frac{\hat{w}}{M_n} - 1)^2}{2\kappa} \le \frac{M_n\kappa}{2}. \tag{8.35}$$

Case 3: $\hat{w} \ge (1+\kappa)M_n$. It turns out that

$$\left\|H_n(\hat{w})\right\| = \left\|\hat{w} - M_n\left(1 + \frac{\kappa}{2}\right)\right\|$$

$$\le \hat{w} - w + \overline{M}_n + \frac{M_n\kappa}{2} = \|e_{n+1}\| + \overline{M}_n + \frac{M_n\kappa}{2}, \tag{8.36}$$

where $\overline{M}_n = \sup_{t \in [0, t_2]}\|w - M_n\|$.

Therefore, it can be obtained that

$$\left\|H_n(\hat{w})\right\| \le \|e_{n+1}\| + \overline{M}_n + \frac{M_n\kappa}{2}, \ \forall\hat{w} \in \mathbb{R}. \tag{8.37}$$

Similarly, for $i = 0, 1, \cdots, n - 1$, it holds that

$$\left\| H_i(\hat{x}_{i+1} - x_r^{(i)}) \right\| \leq \kappa^{n-i} \|e_{i+1}\| + \overline{M}_i + \frac{M_i \kappa}{2}, \ \forall \hat{x}_{i+1} \in \mathbb{R}, \tag{8.38}$$

where $\overline{M}_i = \sup_{t \in [0, t_2]} \|x_{i+1} - x_r^{(i)} - M_i\|$.

Within the time interval $[0, t_2]$, based on Assumptions 8.1–8.4, (8.31), (8.32), (8.37), (8.38), and the boundedness of $s, x^{(0 \sim n-1)}, x_r^{(0 \sim n)}$, there exist positive constants N_0 and N_1 such that

$$\|\hbar\| \leq N_0 + \ell N_1 \|e\|, \ \forall t \in [0, t_2]. \tag{8.39}$$

The derivative of $V_1(e), t \in [0, t_2]$ is

$$\begin{aligned}
\frac{dV_1}{dt}(e) &= \sum_{i=1}^{n+1} \dot{e}_i \frac{\partial V_1}{\partial e_i}(e) \\
&= \frac{1}{\kappa} \left[\sum_{i=1}^{n} (e_{i+1} - h_i(e_1)) \frac{\partial V_1}{\partial e_i}(e) - h_{n+1}(e_1) \frac{\partial V_1}{\partial e_{n+1}}(e) \right] + \hbar \frac{\partial V_1}{\partial e_{n+1}}(e) \\
&\leq -\frac{W_1(e)}{\kappa} + \beta_1 \|e\| (N_0 + \ell N_1 \|e\|) \\
&\leq -\left(\frac{a_3}{\kappa} - \ell \beta_1 N_1 \right) \|e\|^2 + \beta_1 N_0 \|e\| \\
&\leq -\left(\frac{a_3/\kappa - \ell \beta_1 N_1}{a_2} \right) V_1(e) + \frac{\beta_1 N_0}{\sqrt{a_1}} \sqrt{V_1(e)}.
\end{aligned} \tag{8.40}$$

Let $\ell = \frac{\varsigma}{\sqrt{\kappa}}$, where $\varsigma > 0$ satisfies $\kappa < \frac{a_3}{\ell \beta_1 N_1}$. It turns out that

$$\frac{d}{dt} \sqrt{V_1(e)} \leq -\left(\frac{a_3/\kappa - \ell \beta_1 N_1}{2a_2} \right) \sqrt{V_1(e)} + \frac{\beta_1 N_0}{2\sqrt{a_1}}. \tag{8.41}$$

Further, it can be calculated that

$$\|e\| \leq \frac{1}{\sqrt{a_1}} \left(e^{-\left(\frac{a_3/\kappa - \ell \beta_1 N_1}{2a_2} \right)t} \sqrt{V_1(e(0))} + \frac{\beta_1 N_0}{2\sqrt{a_1}} \int_0^t e^{-\left(\frac{a_3/\kappa - \ell \beta_1 N_1}{2a_2} \right)(t-\tau)} d\tau \right). \tag{8.42}$$

This means when $\kappa \to 0$ or $\ell \to \infty$, $\|e\| \to 0, \forall t \in [t_1, t_2]$. Namely, there exists a positive constant ℓ_1 such that for $\forall \ell > \ell_1, t \in [t_1, t_2]$, $\|e\| = o(1/\ell)$, where o represents equivalent infinitesimal.

Since

$$\begin{cases} \|\hat{w}\| \leq \|e_{n+1}\| + \|w\|, \\ \|\hat{x}_{i+1} - x_r^{(i)}\| \leq \kappa^{n-i} \|e_{i+1}\| + \|\eta_{i+1}\|, i = 0, 1, \cdots, n - 1, \end{cases} \tag{8.43}$$

the parameters M_i satisfying

$$
\begin{cases}
M_i > \sup_{[s,\eta^\mathsf{T}]^\mathsf{T} \in \Omega_1} \|\eta_{i+1}\|, i = 0, 1, \cdots, n-1, \\
M_n > \sup_{[s,\eta^\mathsf{T}]^\mathsf{T} \in \Omega_1} \|w\|
\end{cases}
\tag{8.44}
$$

will prevent the saturation phenomenon of $\psi_\kappa(\cdot)$. Consider the following equations:

$$
\frac{dV_2}{dt}(s) = 2s \left[\dot{w} + \sum_{i=0}^{n-1} A_i \left(x^{(i+1)} - x_r^{(i+1)} \right) - x_r^{(n+1)} \right],
\tag{8.45}
$$

$$
\dot{u} \frac{\partial F}{\partial u} = -\ell \left\| \frac{\partial F}{\partial u} \right\| \left[\hat{w} - w + s + \sum_{i=0}^{n-1} A_i \left(\hat{x}_{i+1} - x_{i+1} \right) \right].
\tag{8.46}
$$

Based on the above equation, the derivative of $V_2(s)$ with respect to t over the interval $[t_1, t_2]$ is calculated as

$$
\begin{aligned}
\frac{dV_2}{dt}(s) = 2s &\left[\frac{\partial F}{\partial t} + \sum_{i=1}^{n-1} x_{i+1} \frac{\partial F}{\partial x_i} + \left(\frac{\partial F}{\partial \xi} \right)^\mathsf{T} F_0 + \dot{d} \frac{\partial F}{\partial d} \right. \\
&\left. - \left(\sum_{i=0}^{n-1} A_i \left(x_{i+1} - x_r^{(i)} \right) - x_r^{(n)} \right) \frac{\partial F}{\partial x_n} \right] + 2s^2 \frac{\partial F}{\partial x_n} \\
&- 2\ell s \left[\hat{w} - w + \sum_{i=0}^{n-1} A_i \left(\hat{x}_{i+1} - x_{i+1} \right) \right] \left\| \frac{\partial F}{\partial u} \right\| \\
&- 2\ell s^2 \left\| \frac{\partial F}{\partial u} \right\| + 2s \left[\sum_{i=0}^{n-1} A_i \left(x^{(i+1)} - x_r^{(i+1)} \right) - x_r^{(n+1)} \right].
\end{aligned}
\tag{8.47}
$$

Within the time interval $[t_1, t_2]$, based on Assumptions 8.1–8.4, (8.28) and the boundedness of $x^{(0 \sim n)}$, $x_r^{(0 \sim n+1)}$, there exist positive constants N_2, N_3, N_4 such that for $\forall \ell > \ell_1$, it holds that

$$
\frac{dV_2}{dt}(s) \leq (-\ell N_2 + N_3) s^2 + N_4 \|s\|.
\tag{8.48}
$$

Morover,

$$
\|s\| \leq e^{\left(\frac{-\ell N_2 + N_3}{2}\right)(t-t_1)} \|s(t_1)\| + N_4 \int_{t_1}^{t} e^{\left(\frac{-\ell N_2 + N_3}{2}\right)(t-\tau)} d\tau.
\tag{8.49}
$$

Therefore, there exists a positive constant $\ell_2 > \max\{\ell_1, N_3/N_2\}$ such that for all $\ell > \ell_2, t \in [t_1, t_2]$, $\|s\| = o(1/\ell)$. The derivative of η is calculated as

$$\begin{cases} \dot{\eta}_1 = \eta_2, i \in [n], \\ \dot{\eta}_n = s - A_{0 \sim n-1}\eta. \end{cases} \tag{8.50}$$

And the derivative of $V_3(\eta), t \in [t_1, t_2]$ satisfies

$$\frac{dV_3}{dt}(\eta) = \eta^{\mathsf{T}}(P\Phi_A + \Phi_A^{\mathsf{T}}P)\eta + s\frac{\partial V_3}{\partial \eta_n}(\eta)$$

$$\leq -\|\eta\|^2 + 2\lambda_{\max}(P)\|s\|\|\eta\|. \tag{8.51}$$

Combining (8.48), (8.49) and (8.51), it can be obtained that there exists $\ell_3 > \ell_2$ such that for $\forall \ell > \ell_3, t \in [t_1, t_2]$, it holds that

$$\frac{dV_2}{dt}(s) + \frac{dV_3}{dt}(\eta) < 0. \tag{8.52}$$

This contradicts (8.28). Therefore, there exists a positive constant $\ell^* > 1$ such that for all $\ell > \ell^*, t \geq 0, [s(t), \eta(t)^{\mathsf{T}}]^{\mathsf{T}} \in \Omega_1$. The proof of Lemma 8.1 is completed.

Theorem 8.1 *For the input-nonaffine FAS model (8.1) under the observer (8.13) and the controller dynamics (8.19), if Assumptions 8.1–8.4 hold, then there exist positive constants $\aleph_1, \aleph_2, \aleph_3, t_*$ such that for $\forall [s(0), \eta(0)^{\mathsf{T}}]^{\mathsf{T}} \in \mathbb{R}^{n+1}$ and $t \geq t_*$, it holds that*

$$\begin{cases} \|x_i - \hat{x}_i\| \leq \kappa^{n+2-i}\aleph_1, i \in [n], \\ \|w - \hat{w}\| \leq \kappa\aleph_1, \\ \|s\| \leq \aleph_2/\ell, \\ \|x_i - x_r^{(i-1)}\| \leq \aleph_3/\ell, i \in [n]. \end{cases} \tag{8.53}$$

In particular, a sufficiently large ℓ and a sufficiently small κ can ensure that the observation and tracking error converge to a tiny neighborhood around the origin.

Proof Lemma 8.1 states that for $\forall \ell > \ell^*, t \geq 0$, it follows that

$$[s(t), \eta(t)^{\mathsf{T}}]^{\mathsf{T}} \in \Omega_1. \tag{8.54}$$

Therefore, (8.42), (8.49) and (8.51) still hold for $t \geq 0$. The meaning of (8.42) is that there exist positive constants $\aleph_1$ and T_0 such that for $t \geq T_0, \|e\| \leq \kappa\aleph_1$. Then for $t \geq T_0$, it turns out that

$$\|x_i - \hat{x}_i\| = \kappa^{n+1-i}\|e_i\| \leq \kappa^{n+1-i}\|e\| \leq \kappa^{n+2-i}\aleph_1, i \in [n]. \tag{8.55}$$

Similarly, for $t \geq T_0, \|w - \hat{w}\| \leq \kappa\aleph_1$. Equation (8.49) indicates that there exist positive constants $\aleph_2$ and T_1 such that for $t \geq T_1, \|s\| \leq \aleph_2/\ell$. Therefore, for

$t \geq T_1$, it turns out that

$$\frac{dV_3}{dt}(\eta) \leq -\|\eta\|^2 + \frac{2\lambda_{\max}(P)\aleph_2}{\ell}\|\eta\|$$

$$\leq -\frac{1}{\lambda_{\max}(P)}V_3(\eta) + \frac{2\lambda_{\max}(P)\aleph_2}{\ell\sqrt{\lambda_{\min}(P)}}\sqrt{V_3(\eta)}. \tag{8.56}$$

For $t \geq T_1$, it follows that

$$\|\eta\| \leq \frac{1}{\sqrt{\lambda_{\min}(P)}}\left(e^{-\frac{1}{2\lambda_{\max}(P)}(t-T_1)}\sqrt{V_3(\eta(T_1))}\right.$$

$$\left. + \frac{2\lambda_{\max}(P)\aleph_2}{\ell\sqrt{\lambda_{\min}(P)}}\int_{T_1}^{t} e^{-\frac{1}{2\lambda_{\max}(P)}(t-\tau)}d\tau\right). \tag{8.57}$$

As a result, there exist positive constants $\aleph_3$ and T_2 such that for $t \geq T_2$, it turns out that

$$\|x_i - x_r^{(i-1)}\| \leq \|\eta\| \leq \frac{\aleph_3}{\ell}, i \in [n]. \tag{8.58}$$

After choosing $t_* = \max\{T_0, T_1, T_2\}$, the proof of Theorem 8.1 is completed.

Corollary 8.1 *Consider the input-nonaffine FAS model (8.1) under the observer and controller dynamics*

$$\begin{cases} \dot{\hat{x}}_i = \hat{x}_{i+1} + \frac{b_i}{\kappa^i}(y - \hat{x}_1), i \in [n-1], \\ \dot{\hat{x}}_n = \hat{w} + \frac{b_n}{\kappa^n}(y - \hat{x}_1), \\ \dot{\hat{w}} = \frac{b_{n+1}}{\kappa^{n+1}}(y - \hat{x}_1), \\ \dot{u} = -\ell \operatorname{sign}\left(\frac{\partial F}{\partial u}\right)\left[M_n h_\kappa\left(\frac{\hat{w}}{M_n}\right) - x_r^{(n)} + \sum_{i=0}^{n-1} A_i M_i h_\kappa\left(\frac{\hat{x}_{i+1} - x_r^{(i)}}{M_i}\right)\right], \end{cases} \tag{8.59}$$

where

$$\Psi = \begin{bmatrix} -b_1 & 1 & 0 & \cdots & 0 \\ -b_2 & 0 & 1 & \cdots & 0 \\ \vdots & & & \ddots & \\ -b_n & 0 & 0 & \cdots & 1 \\ -b_{n+1} & 0 & 0 & \cdots & 0 \end{bmatrix} \tag{8.60}$$

is a Hurwitz matrix. If Assumptions 8.1–8.3 hold, then there exist positive constants $\aleph_1, \aleph_2, \aleph_3, t_*$ such that for all $\forall[s(0), \eta(0)^\mathsf{T}]^\mathsf{T} \in \mathbb{R}^{n+1}$ and $t \geq t_*$, it holds that

$$\begin{cases} \|x_i - \hat{x}_i\| \leq \kappa^{n+2-i}\aleph_1, i \in [n], \\ \|w - \hat{w}\| \leq \kappa\aleph_1, \\ \|s\| \leq \aleph_2/\ell, \\ \|x_i - x_r^{(i-1)}\| \leq \aleph_3/\ell, i \in [n]. \end{cases} \tag{8.61}$$

In particular, a sufficiently large ℓ and a sufficiently small κ can ensure that the observation and tracking error converge to a tiny neighborhood around the origin.

Proof Compared with Theorem 8.1, Corollary 8.1 does not require Assumption 8.4. Namely, if Assumption 8.4 always holds, then this corollary is valid. When the parameterized matrix Ψ is Hurwitz, the following Lyapunov equation has a positive definite solution P_b

$$P_b\Psi + \Psi^\mathsf{T} P_b = -I_{n+1}. \tag{8.62}$$

Consider two Lyapunov functions

$$V_1(\mathfrak{z}) = \mathfrak{z}^\mathsf{T} P_b \mathfrak{z}, \ W_1(\mathfrak{z}) = \mathfrak{z}^\mathsf{T} \mathfrak{z}, \ \forall \mathfrak{z} \in \mathbb{R}^{n+1}, \tag{8.63}$$

it can be analyzed that

$$\lambda_{\min}(P_b)\|\mathfrak{z}\|^2 \leq V_1(\mathfrak{z}) \leq \lambda_{\max}(P_b)\|\mathfrak{z}\|^2, \tag{8.64}$$

$$\sum_{i=1}^{n} (\mathfrak{z}_{i+1} - b_i\mathfrak{z}_1) \frac{\partial V_1}{\partial \mathfrak{z}_i}(\mathfrak{z}) - b_{n+1}\mathfrak{z}_1 \frac{\partial V_1}{\partial \mathfrak{z}_{n+1}}(\mathfrak{z}) = -W_1(\mathfrak{z}), \tag{8.65}$$

$$\left\| \frac{\partial V_1}{\partial \mathfrak{z}_{n+1}}(z) \right\| \leq 2\lambda_{\max}(P_b)\|\mathfrak{z}\|. \tag{8.66}$$

Therefore, Assumption 8.4 holds true. The proof of Corollary 8.1 is completed.

8.4   Robust Fault-Tolerant Stabilization Strategies

In practical applications, dynamic systems are prone to various types of faults such as actuator and sensor faults due to their long and arduous tasks. These faults can severely disrupt the parametric FAS structure, rendering the existing FAS theory ineffective. Consider the following unknown input-nonaffine FAS model with actuator and sensor faults:

$$\begin{cases} x^{(n)} = F(x^{(0\sim n-1)}, \sigma(t)u, \xi, d, t), \\ \dot{\xi} = F_0(x^{(0\sim n-1)}, \xi, d, t), \\ y = \theta(t)x, \end{cases} \tag{8.67}$$

where $x, u, y, d \in \mathbb{R}, \xi \in \mathbb{R}^m$ are the system state, the control input, the measurement output, the process disturbance and the unmodeled dynamics, respectively. $F(\cdot), F_0(\cdot)$ are unknown continuous differentiable functions, $\sigma(t)$ is a multiplicative actuator fault, and $\theta(t)$ is a multiplicative sensor fault. Due to inaccurate system measurement, the nonlinear observer (8.13) cannot provide accurate state observation.

Assumption 8.5 The actuator fault satisfies $\sigma(t) \in (0, 1), \forall t \geq 0$, and the sensor fault satisfies $\theta(t) \in [1 - \theta_1, 1 + \theta_1], \forall t \geq 0$ where θ_1 is a positive constant.

Assumption 8.6 There exists a known bijective function $f(u)$ such that for all $t \geq 0$, it follows that

$$\left\| F(x^{(0 \sim n-1)}, \sigma(t)u, \xi, d, t) - f(u) \right\| \leq \rho_3(x^{(0 \sim n-1)}, \xi, d), \tag{8.68}$$

where $\rho_3(x^{(0 \sim n-1)}, \xi, d)$ is a continuous function.

Remark 8.4 The dynamic model $F(\cdot)$ of many practical systems, such as that of robotic arms and permanent magnet synchronous motors, satisfy Assumption 8.6. The requirements for $\sigma(t)$ are also weaker than the general boundedness requirements [11, 74].

The control objective of this subsection is the design of a robust fault-tolerant stabilization controller for the faulty FAS (8.67). An auxiliary system is designed as

$$\begin{cases} \dot{\mathfrak{x}}_i = \mathfrak{x}_{i+1} - \dfrac{1}{\kappa^i} g_i(\mathfrak{x}_1), i \in [n], \\ \dot{\mathfrak{x}}_n = f(u) - \dfrac{1}{\kappa^n} g_n(\mathfrak{x}_1), \end{cases} \tag{8.69}$$

where $\mathfrak{x}_i$ represents the state of the auxiliary system, and $g_i(\cdot), i = 1, 2, \cdots, n$ are the functions to be designed, which are odd functions satisfying $g_i(r) \leq \gamma \| r \|, \gamma > 0$.

Assumption 8.7 There exists positive constants $a_5, a_6, a_7, a_8, \beta_2$ and two positive definite functions $V_4(z), W_2(z), \forall z = [z_1, z_2, \cdots, z_n]^\mathsf{T} \in \mathbb{R}^n$ such that

1. $a_5 \| z \|^2 \leq V_4(z) \leq a_6 \| z \|^2, a_7 \| z \|^2 \leq W_2(z) \leq a_8 \| z \|^2$.
2. $\displaystyle\sum_{i=1}^{n-1} (z_{i+1} - g_i(z_1)) \frac{\partial V_4}{\partial z_i}(z) - g_n(z_1) \frac{\partial V_4}{\partial z_n}(z) \leq -W_2(z)$.
3. $\left\| \dfrac{\partial V_4}{\partial z_i}(z) \right\| \leq \beta_2 \| z \|, i \in [n]$.

Define the following two coordinate transformations

$$\varepsilon = [\varepsilon_1, \varepsilon_2, \cdots, \varepsilon_n]^\mathsf{T}$$
$$= [x_1 - \mathfrak{x}_1, \kappa(x_2 - \mathfrak{x}_2), \cdots, \kappa^{n-1}(x_n - \mathfrak{x}_n)]^\mathsf{T}, \tag{8.70}$$

$$\zeta = [\zeta_1, \zeta_2, \cdots, \zeta_n]^{\mathsf{T}} = [x_1, \kappa\iota\mathfrak{x}_2, \cdots, (\kappa\iota)^{n-1}\mathfrak{x}_n]^{\mathsf{T}}, \tag{8.71}$$

where ι is a constant within the interval $(0, 1)$.

The derivative of ε is

$$\begin{cases} \kappa\dot{\varepsilon}_i = \varepsilon_{i+1} + g_i(\zeta_1 - \varepsilon_1), i \in [n-1], \\ \kappa\dot{\varepsilon}_n = \kappa^n[F(x^{(0\sim n-1)}, \sigma(t)u, \xi, d, t) - f(u)] + g_n(\zeta_1 - \varepsilon_1), \end{cases} \tag{8.72}$$

and the derivative of ζ is

$$\begin{cases} \dot{\zeta}_1 = \dfrac{\zeta_2}{\kappa\iota} + \dfrac{\varepsilon_2}{\kappa}, \\ \dot{\zeta}_2 = \dfrac{\zeta_3}{\kappa\iota} - \dfrac{\iota}{\kappa}g_2(\zeta_1 - \varepsilon_1), \\ \quad\vdots \\ \dot{\zeta}_n = \dfrac{(\kappa\iota)^n f(u)}{\kappa\iota} - \dfrac{\iota^{n-1}}{\kappa}g_n(\zeta_1 - \varepsilon_1). \end{cases} \tag{8.73}$$

Taking into account the inaccuracy of the measurement output, the following robust fault-tolerant stabilization controller is designed as

$$\begin{aligned} u &= f^{-1}\left(-\frac{A_{0\sim n-1}(y, \zeta_2, \zeta_3, \cdots, \zeta_n)}{(\kappa\iota)^n}\right) \\ &= f^{-1}\left(-\frac{A_{0\sim n-1}(y, \kappa\iota\mathfrak{x}_2, (\kappa\iota)^2\mathfrak{x}_3, \cdots, (\kappa\iota)^{n-1}\mathfrak{x}_n)}{(\kappa\iota)^n}\right), \end{aligned} \tag{8.74}$$

where $f^{-1}(\cdot)$ represents the inverse function of $f(\cdot)$.

Define

$$\varpi_2 = \sup_{[\varepsilon^{\mathsf{T}}, \zeta^{\mathsf{T}}]^{\mathsf{T}} \in \mathbb{R}^{2n}, \|[\varepsilon^{\mathsf{T}}, \zeta^{\mathsf{T}}]^{\mathsf{T}}\| \leq \|[\varepsilon(0)^{\mathsf{T}}, \zeta(0)^{\mathsf{T}}]^{\mathsf{T}}\|} (V_4(\varepsilon) + V_5(\zeta)) + 1, \tag{8.75}$$

$$V_5(\zeta) = \zeta^{\mathsf{T}} P \zeta, \tag{8.76}$$

and two compact sets

$$\Omega_2 = \{[\varepsilon^{\mathsf{T}}, \zeta^{\mathsf{T}}]^{\mathsf{T}} \mid [\varepsilon^{\mathsf{T}}, \zeta^{\mathsf{T}}]^{\mathsf{T}} \in \mathbb{R}^{2n}, V_4(\varepsilon) + V_5(\zeta) \leq \varpi_2\}, \tag{8.77}$$

$$\Omega_3 = \{[\varepsilon^{\mathsf{T}}, \zeta^{\mathsf{T}}]^{\mathsf{T}} \mid [\varepsilon^{\mathsf{T}}, \zeta^{\mathsf{T}}]^{\mathsf{T}} \in \mathbb{R}^{2n}, V_4(\varepsilon) + V_5(\zeta) \leq \varpi_2 + 1\}. \tag{8.78}$$

Lemma 8.2 *For the faulty FAS model* (8.67) *under the controller* (8.74), *if Assumptions 8.1–8.3, 8.5–8.7 ***and* $2\|P\|A_0 < 1/\theta_1$ *hold, then there exists a positive constant* $\kappa^\star$ *such that for* $\forall\kappa \in (0, \kappa^\star)$ *and* $t \geq 0$, *it turns out that* $[\varepsilon(t)^{\mathsf{T}}, \zeta(t)^{\mathsf{T}}]^{\mathsf{T}} \in \Omega_3$.

Proof Based on $V_4(\varepsilon(0)) + V_5(\zeta(0)) < \varpi_2$ and the continuity of $\varepsilon(t)$, $\zeta(t)$, it is obtained that

$$[\varepsilon(0)^\mathsf{T}, \zeta(0)^\mathsf{T}]^\mathsf{T} \in \Omega_2 - \partial\Omega_2, \tag{8.79}$$

then there exists $t_3 > 0$ such that $\forall t \in [0, t_3]$, $[\varepsilon(t)^\mathsf{T}, \zeta(t)^\mathsf{T}]^\mathsf{T} \in \Omega_2 - \partial\Omega_2$. The remaining proof is completed by contradiction. Assuming that this lemma does not hold, i.e., for any $\check{\kappa} > 0$, there exist $\kappa \in (0, \check{\kappa})$ and $t \geq 0$ such that

$$[\varepsilon(t)^\mathsf{T}, \zeta(t)^\mathsf{T}]^\mathsf{T} \in \mathbb{R}^{2n} - \Omega_3. \tag{8.80}$$

$[\varepsilon(t)^\mathsf{T}, \zeta(t)^\mathsf{T}]^\mathsf{T}$ is a continuous function, and for $\forall t \in [0, t_3]$, it holds that

$$[\varepsilon(t)^\mathsf{T}, \zeta(t)^\mathsf{T}]^\mathsf{T} \in \Omega_2 - \partial\Omega_2. \tag{8.81}$$

Therefore, there exist $t_5 > t_4 > t_3$ such that

$$\begin{cases} [\varepsilon(t_4)^\mathsf{T}, \zeta(t_4)^\mathsf{T}]^\mathsf{T} \in \partial\Omega_2, \\ [\varepsilon(t_5)^\mathsf{T}, \zeta(t_5)^\mathsf{T}]^\mathsf{T} \in \Omega_3 - \Omega_2, \\ V_4(\varepsilon(t_5)) + V_5(\zeta(t_5)) > \varpi_2 + 0.5, \\ [\varepsilon(t)^\mathsf{T}, \zeta(t)^\mathsf{T}]^\mathsf{T} \in \Omega_3 - \Omega_2^\circ, \forall t \in [t_4, t_5], \\ [\varepsilon(t)^\mathsf{T}, \zeta(t)^\mathsf{T}]^\mathsf{T} \in \Omega_3, \forall t \in [0, t_5]. \end{cases} \tag{8.82}$$

The derivative of $V_4(\varepsilon)$, $t \in [0, t_5]$ is

$$\begin{aligned}
\frac{dV_4}{dt}(\varepsilon) &= \sum_{i=1}^{n} \dot{\varepsilon}_i \frac{\partial V_4}{\partial \varepsilon_i}(\varepsilon) \\
&= \frac{1}{\kappa}\left[\sum_{i=1}^{n-1}(\varepsilon_{i+1} - g_i(\varepsilon_1))\frac{\partial V_4}{\partial \varepsilon_i}(\varepsilon) - g_n(\varepsilon_1)\frac{\partial V_4}{\partial \varepsilon_n}(\varepsilon)\right] \\
&\quad + \kappa^{n-1}\left[F(x^{(0 \sim n-1)}, \sigma(t)u, \xi, d, t) - f(u)\right]\frac{\partial V_4}{\partial \varepsilon_n}(\varepsilon) \\
&\quad + \frac{1}{\kappa}\sum_{i=1}^{n}[g_i(\varepsilon_1) - g_i(\varepsilon_1 - \zeta_1)]\frac{\partial V_4}{\partial \varepsilon_i}(\varepsilon) \\
&\leq -\frac{W_2(\varepsilon)}{\kappa} + \beta_2\|\varepsilon\|\left[\kappa^{n-1}\rho_3(x^{(0 \sim n-1)}, \xi, d) + \frac{n\gamma}{\kappa}\|\zeta_1\|\right] \\
&\leq -\frac{a_7}{2\kappa}\|\varepsilon\|^2 + \kappa^{n-1}\beta_2 N_5\|\varepsilon\| + \frac{n^2\beta_2^2\gamma^2}{2a_7\kappa}\|\zeta_1\|^2, \tag{8.83}
\end{aligned}$$

where N_5 is a constant based on Assumption 8.6 within the time interval $[t_0, t_5]$. The derivative of $V_5(\zeta), t \in [t_4, t_5]$ is calculated as

$$\frac{dV_5}{dt}(\zeta) = -\frac{1}{\kappa\iota}\|\zeta\|^2 + \frac{2}{\kappa\iota}\zeta^\mathsf{T} P B_n A_0 (1 - \theta(t))\zeta_1$$
$$+ 2\zeta^\mathsf{T} P\left[\frac{\iota}{\kappa}\mathscr{G}(\varepsilon_1 - \zeta_1) + \frac{1}{\kappa}C_n^\mathsf{T}\varepsilon_2\right], \qquad (8.84)$$

where

$$\mathscr{G} = \left[0, g_2(\varepsilon_1 - \zeta_1), \iota g_3(\varepsilon_1 - \zeta_1), \cdots, \iota^{n-2}g_n(\varepsilon_1 - \zeta_1)\right]^\mathsf{T}. \qquad (8.85)$$

Further,

$$\frac{dV_5}{dt}(\zeta) \leq -\frac{1 - 2\|P\|A_0(1 - \theta(t))}{\kappa\iota}\|\zeta\|^2$$
$$+ \frac{2\|P\|\iota\gamma}{\kappa}\|\zeta\|^2 + \frac{\|P\|(1 + \iota\gamma)}{\kappa}(\|\zeta\|^2 + \|\varepsilon\|^2). \qquad (8.86)$$

If $2\|P\|A_0 < 1/\theta_1$, it follows that

$$\varkappa \triangleq 1 - 2\|P\|A_0(1 - \theta(t)) > 0 \qquad (8.87)$$

and

$$\frac{dV_4}{dt}(\varepsilon) + \frac{dV_5}{dt}(\zeta) \leq -\frac{\varkappa - N_6}{\kappa\iota}\|\zeta\|^2 - \frac{a_7 - N_7}{2\kappa}\|\varepsilon\|^2 + \kappa^{n-1}\beta_2 N_5\|\varepsilon\|, \qquad (8.88)$$

where

$$N_6 = 2\|P\|\iota^2\gamma + \|P\|\iota(1 + \iota\gamma) + \frac{\iota n^2\beta_2^2\gamma^2}{2a_7}, \qquad (8.89)$$

$$N_7 = 2\|P\|(1 + \iota\gamma). \qquad (8.90)$$

If the parameters satisfy

$$1 - 2\|P\|A_0(1 - \theta_1) > N_6, a_7 > N_7, \qquad (8.91)$$

it can be obtained that

$$\|[\varepsilon^\mathsf{T}, \zeta^\mathsf{T}]^\mathsf{T}\| \leq \frac{\kappa^n \beta_2 N_5}{\min\{(\varkappa - N_6)/\iota, (a_7 - N_7)/2\}}. \qquad (8.92)$$

This indicates that as $\kappa \to 0$, $\|[\varepsilon^\mathsf{T}, \zeta^\mathsf{T}]^\mathsf{T}\| \to 0$ for $t \in [t_4, t_5]$, i.e., there exists a positive constant κ_1 such that for all $\kappa \in (0, \kappa_1)$ and $t \in [t_4, t_5]$, it holds that

$\|[\varepsilon^{\mathsf{T}}, \zeta^{\mathsf{T}}]^{\mathsf{T}}\| = o(\kappa)$. Therefore, for $\forall \kappa \in (0, \kappa_1)$ and $t \in [t_4, t_5]$, it turns out that

$$\frac{dV_4}{dt}(\varepsilon) + \frac{dV_5}{dt}(\zeta) < 0. \tag{8.93}$$

This contradicts (8.82). Therefore, there exists a positive constant $\kappa^\star$ such that for $\forall \kappa \in (0, \kappa^\star)$ and $t \geq 0$, it holds that $[\varepsilon(t)^{\mathsf{T}}, \zeta(t)^{\mathsf{T}}]^{\mathsf{T}} \in \Omega_3$. The proof of Lemma 8.2 is completed.

Theorem 8.2 *For the faulty FAS* (8.67) *under the controller* (8.74), *if Assumptions 8.1–8.3, 8.5–8.7* ***and* $2\|P\|A_0 < 1/\theta_1$ *hold, then there exist positive constants* $\aleph_4$, $\aleph_5$, $t^\star$ *such that for* $\forall [\varepsilon(0)^{\mathsf{T}}, \zeta(0)^{\mathsf{T}}]^{\mathsf{T}} \in \mathbb{R}^{2n}$ *and* $t \geq t^\star$, *the following conclusions hold:*

$$\begin{cases} \|[\varepsilon(t)^{\mathsf{T}}, \zeta(t)^{\mathsf{T}}]^{\mathsf{T}}\| \leq \kappa^n \aleph_4, \\ \|x_i\| \leq \kappa^{n+1-i} \aleph_5, i \in [n]. \end{cases} \tag{8.94}$$

In particular, a sufficiently small κ can ensure that the stabilization error converges to a tiny neighborhood around the origin.

Proof Lemma 8.2 states that $[\varepsilon(t)^{\mathsf{T}}, \zeta(t)^{\mathsf{T}}]^{\mathsf{T}} \in \Omega_3, \forall \kappa \in (0, \kappa^\star), t \geq 0$. Therefore, (8.92) is applicable when $t \geq 0$. Then there exist positive constants $\aleph_4$ and $t^\star$ such that for $t \geq t^\star$, it holds that $\|[\varepsilon(t)^{\mathsf{T}}, \zeta(t)^{\mathsf{T}}]^{\mathsf{T}}\| \leq \kappa^n \aleph_4$. Additionally,

$$x_1 = \zeta_1, x_i = \frac{1}{\kappa^{i-1}}\varepsilon_i + \frac{1}{(\kappa\iota)^{i-1}}\zeta_i, i = 2, 3, \cdots, n. \tag{8.95}$$

Therefore, there exists a positive constant $\aleph_5$ such that for $t \geq t^\star$, it holds that $\|x_i\| \leq \kappa^{n+1-i}\aleph_5, i \in [n]$. Theorem 8.2 has been proved.

Corollary 8.2 *Consider the faulty FAS model* (8.67) *under the observer and controller*

$$\begin{cases} \dot{\mathfrak{x}}_i = \hat{x}_{i+1} - \dfrac{b_i}{\kappa^i}\mathfrak{x}_i, i \in [n-1], \\ \dot{\mathfrak{x}}_n = f(u) - \dfrac{b_n}{\kappa^n}\mathfrak{x}_1, \\ u = f^{-1}\left(\dfrac{-A_{0\sim n-1}(y, \kappa\iota\mathfrak{x}_2, (\kappa\iota)^2\mathfrak{x}_3, \cdots, (\kappa\iota)^{n-1}\mathfrak{x}_n)}{(\kappa\iota)^n}\right), \end{cases} \tag{8.96}$$

where

$$\Psi_b = \begin{bmatrix} -b_1 & 1 & 0 & \cdots & 0 \\ -b_2 & 0 & 1 & \cdots & 0 \\ \vdots & & & \ddots & \\ -b_{n-1} & 0 & 0 & \cdots & 1 \\ -b_n & 0 & 0 & \cdots & 0 \end{bmatrix} \tag{8.97}$$

is a Hurwitz matrix. If Assumptions 8.1–8.3, 8.5–8.6 and $2\|P\|A_0 < 1/\theta_1$ hold, then there exist positive constants $\aleph_4, \aleph_5, t^\star$ such that for $\forall[\varepsilon(0)^\mathsf{T}, \zeta(0)^\mathsf{T}]^\mathsf{T} \in \mathbb{R}^{2n}$ and $t \geq t^\star$, the following conclusions hold:

$$\begin{cases} \|[\varepsilon(t)^\mathsf{T}, \zeta(t)^\mathsf{T}]^\mathsf{T}\| \leq \kappa^n \aleph_4, \\ \|x_i\| \leq \kappa^{n+1-i} \aleph_5, i \in [n]. \end{cases} \tag{8.98}$$

In particular, a sufficiently small κ can ensure that the stabilization error converges to a tiny neighborhood around the origin.

Proof If Assumption 8.7 holds true, then this conclusion is valid. The proof can be found in Corollary 8.1.

Remark 8.5 In the unknown input-nonaffine FAS model, the proposed robust stabilization controller can effectively cope with multiplicative actuator faults and multiplicative sensor faults. Different from the traditional FAS controllers based on observers, this chapter proposes an FAS controller based on an auxiliary system to mitigate the effects from two types of faults. The auxiliary system can approximate the practical controlled system by adjusting two parameters κ and ι. Theorem 8.2 and Corollary 8.2 prove that two appropriate parameters κ and ι can reduce the negative impacts of faults. However, this approach also has some drawbacks. For example, this robust controller cannot handle additive sensor faults, and it cannot complete the trajectory tracking task in the case of additive sensor faults.

8.5 Simulation Studies

8.5.1 A Tracking Control Experiment

Consider the following unknown input-nonffine FAS model:

$$\begin{cases} x^{(3)} = F(x^{(0\sim2)}, u, \xi, d, t), \\ \dot{\xi} = F_0(x^{(0\sim2)}, u, \xi, d, t), \\ y = x, \end{cases} \tag{8.99}$$

where

$$F = (1 + \sin \dot{x}^2)(x - \ddot{x}) + \frac{u + u^3}{1 + x^2} + \sin \xi + 2d, \tag{8.100}$$

$$F_0 = -(x^2 + d^2) \sin \xi, d = 0.1 \sin t. \tag{8.101}$$

In this experiment, F, F_0 and d are unknown, and the sign of $\frac{\partial F}{\partial u}$ is known, i.e., $\text{sign}(\frac{\partial F}{\partial u}) = 1$. It is easy to verify that the FAS model (8.99) complies with

Assumptions 8.1–8.3. The reference signal is $x_r = 0.5 \sin t$, and the observer is designed as

$$
\begin{cases}
\dot{\hat{x}}_1 = \hat{x}_2 + \dfrac{4}{\kappa}(y - \hat{x}_1) + \kappa^2 \phi \left(\dfrac{y - \hat{x}_1}{\kappa^3} \right), \\
\dot{\hat{x}}_2 = \hat{x}_3 + \dfrac{6}{\kappa^2}(y - \hat{x}_1), \\
\dot{\hat{x}}_3 = \hat{w} + \dfrac{4}{\kappa^3}(y - \hat{x}_1), \\
\dot{\hat{w}} = \dfrac{1}{\kappa^4}(y - \hat{x}_1),
\end{cases}
\tag{8.102}
$$

where

$$
\phi(r) =
\begin{cases}
-\dfrac{1}{20}, & r \le -\dfrac{\pi}{2}, \\
\dfrac{1}{20} \sin r, & -\dfrac{\pi}{2} \le r \le \dfrac{\pi}{2}, \\
\dfrac{1}{20}, & r \ge \dfrac{\pi}{2}.
\end{cases}
\tag{8.103}
$$

In the observer, the nonlinear functions are set as $h_1(z) = 4z_1 + \phi(z_1)$, $h_2(z) = 6z_1$, $h_3(z) = 4z_1$, and $h_4(z) = z_1$, which satisfy Assumption 8.4 [43]. Then the controller dynamics is designed as

$$
\dot{u} = -\ell \left[M_3 \psi_\kappa \left(\frac{\hat{w}}{M_3} \right) - x_r^{(3)} + \sum_{i=0}^{2} A_i M_i \psi_\kappa \left(\frac{\hat{x}_{i+1} - x_r^{(i)}}{M_i} \right) \right].
\tag{8.104}
$$

The other parameters are set as $\kappa = 0.01$, $\ell = 12.5$, $A_0 = 2$, $A_1 = 4$, $A_2 = 3$, $M_0 = 2$, $M_1 = 1$, $M_2 = 3$, $M_3 = 6$. The experimental results are shown in Figs. 8.1 and 8.2. Figure 8.1 indicates that the proposed observer and controller can complete the trajectory tracking task of the unknown FAS. In Fig. 8.1a, there is still a small tracking error, and the reason for this error is that the parameter ℓ cannot be set to infinity. Figure 8.2 shows that the tracking controller signal is smooth and has a good application prospect in the unknown system.

8.5.2 A Robust Fault-Tolerant Stabilization Experiment

Consider a single-link robot model:

$$
\begin{cases}
\mathcal{M} \ddot{q} + G(q) + T_\xi + T_d = [1 + T_\tau(q, \dot{q}, \tau_q)] \sigma(t) \tau_q, \\
y = \theta(t) q,
\end{cases}
\tag{8.105}
$$

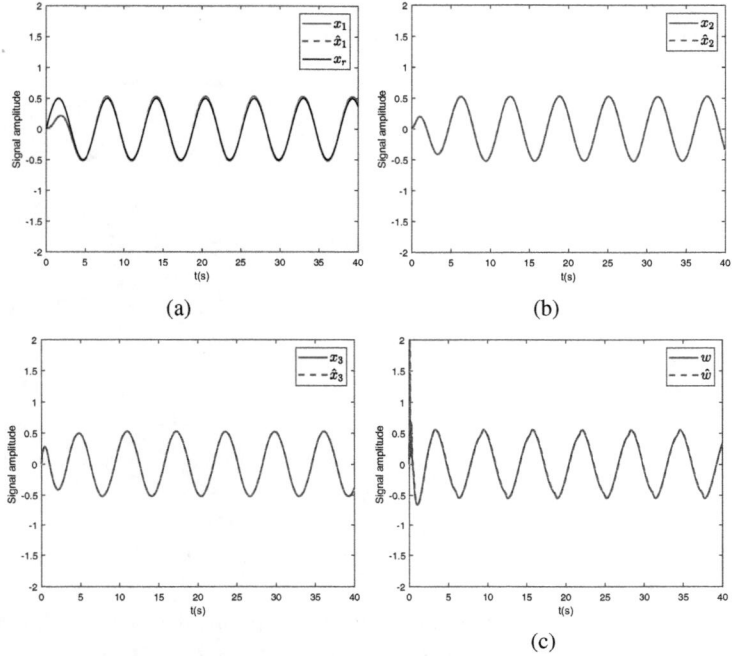

(a)

(b)

(c)

Fig. 8.1 The signal trajectories in a tracking control experiment

Fig. 8.2 The control input in a tracking control experiment

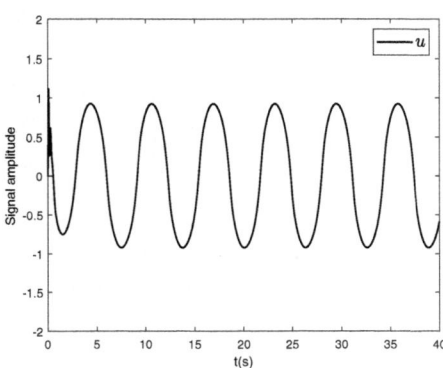

where, $q, \tau_q, T_\xi, T_d, T_\tau(q, \dot{q}, \tau_q), \sigma(t), \theta(t)$ represent the angle, the input torque, the unmodeled dynamics, the process disturbance, the input perturbation, the actuator fault, and the sensor fault, respectively. The unknown terms and faults are set as

$$\mathcal{M} = 1, G(q) = 10 \sin q, \dot{T}_\xi = -q^2 \cos T_\xi, \tag{8.106}$$

$$T_d = 0.1 \sin t, T_\tau(q, \dot{q}, \tau_q) = \frac{1}{\tau_q^2 + q^2 + 1}, \tag{8.107}$$

Fig. 8.3 The signal
trajectories in a robust
fault-tolerant stabilization
experiment

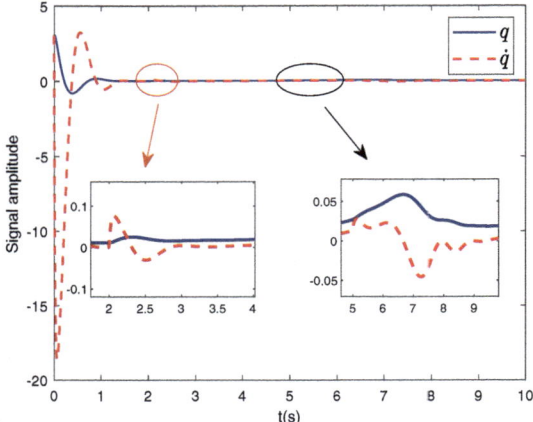

$$\sigma(t) = \begin{cases} 1, 0 \leq t \leq 5\text{s}, \\ 0.5, t > 5\text{s}, \end{cases} \quad \theta(t) = \begin{cases} 1, 0 \leq t \leq 2\text{s}, \\ 0.4 + 0.3\sin t, t > 2\text{s}. \end{cases} \tag{8.108}$$

The robot model (8.105) complies with Assumptions 8.1–8.3 and 8.5–8.6 with $f(u) = u$. The fault-tolerant stabilization controller based on an auxiliary system is designed as

$$\begin{cases} \dot{\mathfrak{x}}_1 = \mathfrak{x}_2 - \dfrac{b_1}{\kappa}\mathfrak{x}_1, \\ \dot{\mathfrak{x}}_2 = u - \dfrac{b_2}{\kappa^2}\mathfrak{x}_1, \\ u = -\dfrac{1}{(\kappa\iota)^2}A_{0\sim 1}(y, \kappa\iota\mathfrak{x}_2). \end{cases} \tag{8.109}$$

The other parameters are set as $\kappa = 0.1$, $\iota = 0.3$, $A_0 = 0.25$, $A_1 = 1$, $b_1 = 5$, $b_2 = 2$, $\mathfrak{x}_1(0) = 0$, and $\mathfrak{x}_2(0) = 0$. The experimental results are shown in Fig. 8.3. This controller can effectively stabilize the robot system (8.105).

8.6   Notes and References

This chapter investigates the trajectory tracking control and fault-tolerant stabilization control problems for unknown input-nonaffine FASs. In the design of the tracking controller, in order to maintain the pre-closed-loop linear structure of the input-nonaffine FASs, an observer-based controller dynamics equation is proposed. In the design of the robust fault-tolerant stabilization controller, an auxiliary system is used to approximate the FAS to handle multiplicative actuator faults and multiplicative sensor faults. The controllers weakly dependent on models have further enriched the FAS theory.

Chapter 9
Conclusion

9.1 Summary

In the first-order state-space theory, most controller structures are complex, and the corresponding fault-tolerant control (FTC) design processes lack universality. For instance, the FTC implemented by methods such as sliding mode control, $\mathcal{H}_\infty$ control, and backstepping control have different controller structures and parameter tuning rules. Fortunately, the high-order fully actuated system (FAS) model is a fully parameterized model, with universal parameter tuning rules, and many original mechanism models of practical systems are FAS models. Nevertheless, the most important nonlinear cancellation principle in the FAS theory is severely disrupted by factors such as faults, making it necessary to innovate the FAS controller structure in different fault conditions. The main content and conclusions of the monograph are as follows:

1. For nonlinear FASs with component faults, process disturbances, and unmodeled dynamics simultaneously, an active fault-tolerant control (AFTC) framework based on dynamic data modeling is proposed. When the full-order state cannot be directly measured, an integrated structure of the observer and controller for the FAS model is established. This parameterized structure can ensure the bounded convergence of the observation error and tracking error. Based on this integrated parameter structure, a closed-loop FAS dynamic data modeling method is designed, and it is proved that the FAS via this data model can approximate the model-driven FAS. Using the dynamic data model, the detectability condition of component faults is given, and finally, the AFTC can achieve the bounded convergence of output tracking error, and its compensation performance for large-amplitude faults is stronger than that of the general passive fault-tolerant control (PFTC).

2. For nonlinear FASs with actuator faults and process disturbances, a robust adaptive fault-tolerant control (RAFTC) framework is proposed. This work

© The Author(s) 2026
D. Zhou, M. Cai, *Fault-Tolerant Control for Fully Actuated Systems*,
https://doi.org/10.1007/978-981-95-0691-0_9

develops an observer-based fault-tolerant controller for multi-input-multi-output tracking error system. For multiplicative actuator faults, a robust FTC strategy is employed, while an adaptive FTC strategy addresses additive faults. By integrating these with the observer-based FAS controller, the high-order error system is ultimately uniformly bounded. The proposed RAFTC approach shows superior performance compared to conventional linear active disturbance rejection control techniques.

3. For nonlinear FASs with actuator faults, process disturbances, time-varying parameters, and high-frequency measurement noise simultaneously, a novel low-power fault-tolerant control (LPFTC) framework is proposed. Compared with the standard FTC, this LPFTC can reduce the observer gain from the standard $n+1$th power to 2nd power. It is proved that the noise suppression performance of the linear LPFTC is better than that of the standard FTC. A four-order simulation case illustrates that the controller energy of the LPFTC is reduced by several orders of magnitude compared to the standard FTC controller.

4. Considering the response speed and fault compensation speed of FASs, a parameterized finite-time fault-tolerant control (FTFTC) framework is proposed based on the principle of homogeneity. Compared with most state-space-based FTFTC structures, the FAS-based FTFTC structure is simple and the parameter tuning is also simple. Compared with asymptotic FTC for FASs, the fault compensation accuracy under the proposed FTFTC is larger, and the fault compensation speed is faster.

5. For nonlinear FASs with sensor faults, a self-healing fault-tolerant control (SHFTC) framework based on redundancy principle is proposed. Most existing sensor FTC methods require hardware redundancy, model redundancy, distributed redundancy, etc., and these assumptions can be unified as the requirement of system observability. To analyze the system observability, the observable space decomposition under each measurement is performed, and then the overall observable space is determined based on the fusion observation. The observability and redundant observability conditions of FASs are given, and the ideal sensor fault tolerance is creatively defined. Then, a necessary and a sufficient condition for sensor fault tolerance is derived. This necessary and sufficient condition has important guiding significance for the number and location of sensors in practical systems. Finally, the reconstruction rules of the FAS controller are designed on the dynamically adjusted fusion observation, and the SHFTC can accommodate sensor faults in both steady-state and transient processes.

6. For nonlinear FASs with actuator faults, sensor faults, and general measurement noise, an AFTC general design is first given. This general design can only guarantee the basic fault-tolerant function and is sensitive to noise. Then, an AFTC saturation design is presented, which not only ensures the bounded convergence of the error system but also can suppress noise to a certain extent. For a linear FAS with impulsive-type noise, it is proved that the noise suppression performance of the AFTC saturation design is superior to that of the general design. In addition, an AFTC dead-zone design is given, which can also ensure

the bounded convergence of the error system. For a linear FAS with constant-type noise, it is proved that the noise suppression performance of the AFTC dead-zone design is superior to that of the general design.

7. For the less-studied input-nonaffine FASs, the parameterization design standard for the FAS controllers is provided. When the dynamics of the input-nonaffine FAS model is unknown, a differential dynamic equation for the control input signal is designed based on this parameterization design standard. It is proved that the ultimately determined controller can numerically approximate the ideal FAS controller. When the input-nonaffine FAS encounters actuator and sensor faults, a robust FTC based on an auxiliary system can ensure the bounded convergence of the error system.

9.2 Future Outlook

Although this monograph has conducted a preliminary exploration of the FTC theory and methods for FASs under various fault types, there are still many issues worthy of attention:

1. The FTC problem for multiple-input-mutiple-output FASs with incomplete measurability. For nonlinear multiple-input-mutiple-output systems and outputs, the FAS theory can fully demonstrate its fundamental advantages. Many state-space methods are difficult to be applied in the analysis and design of multiple-input-mutiple-output systems, while the parameterized FAS structure is more conducive to the analysis and design of these nonlinear systems. Particularly, when there are actuator or sensor faults in these complex systems, how to design the multi-variable parameterized structure is worthy of further study.

2. The FTC problem for FASs with strong disturbances. The term "strong" in strong disturbances is a relative degree, corresponding to whether the capacity of control input can handle the disturbance. The controller structures under the previous state-space approaches are complex and have different capabilities, while the controller structures under the FAS approaches are parameterized. Can the quantitative analysis of limit capability on control input provide a definition for strong disturbances? When the disturbance exceeds the limit capability, faults and disturbances need other handling ideas, and the coupling effects of faults and disturbances deserve discussion.

3. The collaborative design problem of closed-loop fault diagnosis and FTC for FASs. The FAS theory is a closed-loop system theory. The residual indicator for fault diagnosis in closed-loop systems has always lacked a universal residual indicator. Can the residual indicator of FASs have a parameterized structure? Can an index take into account both fault diagnosis performance and FTC performance?

4. The practical application of FAS-based FTC approaches in fields such as robots and permanent magnet synchronous motors.

References

1. Adiguzel, F., Yalcin, Y.: Immersion and invariance disturbance observer-based nonlinear discrete-time control for fully actuated mechanical systems. Int. J. Syst. Sci. **53**(2), 388–401 (2022)
2. Aliakbari, S., Ayati, M., Osman, J.H., Sam, Y.M.: Second-order sliding mode fault-tolerant control of heat recovery steam generator boiler in combined cycle power plants. Appl. Thermal Eng. **50**(1), 1326–1338 (2013)
3. Arefifar, S.A., Mohamed, Y.A.R.I., El-Fouly, T.H.: Comprehensive operational planning framework for self-healing control actions in smart distribution grids. IEEE Trans. Power Syst. **28**(4), 4192–4200 (2013)
4. Astolfi, D., Marconi, L.: A high-gain nonlinear observer with limited gain power. IEEE Trans. Autom. Control **60**(11), 3059–3064 (2015)
5. Astolfi, D., Alessandri, A., Zaccarian, L.: Stubborn and dead-zone redesign for nonlinear observers and filters. IEEE Trans. Autom. Control **66**(2), 667–682 (2021)
6. Badihi, H., Zhang, Y., Pillay, P., Rakheja, S.: Fault-tolerant individual pitch control for load mitigation in wind turbines with actuator faults. IEEE Trans. Ind. Electron. **68**(1), 532–543 (2021)
7. Basheer, A.A., Palanimuthu, K., Lee, S.R., Joo, Y.H.: Efficiency enhancement using fault-tolerant sliding mode control for the PMVG-based WTS under actuator faults. IEEE Trans. Ind. Electron. **71**(1), 513–523 (2024)
8. Benosman, M., Lum, K.Y.: Passive actuators' fault-tolerant control for affine nonlinear systems. IEEE Trans. Control Syst. Technol. **18**(1), 152–163 (2010)
9. Bhat, S., Bernstein, D.: Continuous finite-time stabilization of the translational and rotational double integrators. IEEE Trans. Autom. Control **43**(5), 678–682 (1998)
10. Bhat, S.P., Bernstein, D.S.: Geometric homogeneity with applications to finite-time stability. Math. Control Signals Syst. **17**, 101–127 (2005)
11. Cai, M., He, X., Zhou, D.: Fault-tolerant tracking control for nonlinear observer-extended high-order fully-actuated systems. J. Franklin Inst. **360**(1), 136–153 (2023)
12. Chen, C.C., Xu, S.S.D., Liang, Y.W.: Study of nonlinear integral sliding mode fault-tolerant control. IEEE/ASME Trans. Mech. **21**(2), 1160–1168 (2016)
13. Chen, L., Zhu, Y., Wu, F., Zhao, Y.: Fault estimation observer design for markovian jump systems with nondifferentiable actuator and sensor failures. IEEE Trans. Cybern. **53**(6), 3844–3858 (2023)
14. Chen, S., Wang, W., Fan, J., Ji, Y.: Impact angle constraint guidance law using fully-actuated system approach. Aerospace Sci. Technol. **136**, 108220 (2023)

D. Zhou, M. Cai, *Fault-Tolerant Control for Fully Actuated Systems*,
https://doi.org/10.1007/978-981-95-0691-0

15. Cocetti, M., Tarbouriech, S., Zaccarian, L.: High-gain dead-zone observers for linear and nonlinear plants. IEEE Control Syst. Lett. **3**(2), 356–361 (2019)
16. Colombo, L.J., Giribet, J.I.: Learning-based fault-tolerant control for a hexarotor with model uncertainty. IEEE Trans. Control Syst. Technol. **32**(2), 672–679 (2024)
17. Cristofaro, A., Polycarpou, M.M., Johansen, T.A.: Fault-tolerant control allocation for overactuated nonlinear systems. Asian J. Control **20**(2), 621–634 (2018)
18. Davila, J., Tranninger, M., Fridman, L.: Finite-time state observer for a class of linear time-varying systems with unknown inputs. IEEE Trans. Autom. Control **67**(6), 3149–3156 (2022)
19. Ding, R., Cheng, M., Zheng, S., Xu, B.: Sensor-fault-tolerant operation for the independent metering control system. IEEE/ASME Trans. Mechatron. **26**(5), 2558–2569 (2021)
20. Duan, G.R.: High-order system approaches: I. Fully-actuated systems and parametric designs. Acta Autom. Sin. **46**(7), 1333–1345 (2020)
21. Duan, G.R.: High-order system approaches: II. Controllability and full-actuation. Acta Autom. Sin. **46**(8), 1571–1581 (2020)
22. Duan, G.R.: High-order system approaches: III. Observability and observer design. Acta Autom. Sin. **46**(9), 1885–1895 (2020)
23. Duan, G.: High-order fully actuated system approaches: part I. Models and basic procedure. Int. J. Syst. Sci. **52**(2), 422–435 (2021)
24. Duan, G.: High-order fully actuated system approaches: part II. Generalized strict-feedback systems. Int. J. Syst. Sci. **52**(3), 437–454 (2021)
25. Duan, G.: High-order fully actuated system approaches: part III. Robust control and high-order backstepping. Int. J. Syst. Sci. **52**(5), 952–971 (2021)
26. Duan, G.: High-order fully actuated system approaches: part V. Robust adaptive control. Int. J. Syst. Sci. **52**(10), 2129–2143 (2021)
27. Duan, G.: High-order fully-actuated system approaches: part VI. Disturbance attenuation and decoupling. Int. J. Syst. Sci. **52**(10), 2161–2181 (2021)
28. Duan, G.: High-order fully actuated system approaches: part VII. Controllability, stabilisability and parametric designs. Int. J. Syst. Sci. **52**(14), 3091–3114 (2021)
29. Duan, G.: State space approaches vs fully actuated system approaches (i). All About Syst. Control **8**(2), 33–44 (2021)
30. Duan, G.: Fully actuated system approaches for continuous-time delay systems: part 1. Systems with state delays only. Sci. China Inf. Sci. **66**(1), 112201 (2023)
31. Duan, G.R.: Brockett's second example: an FAS approach treatment. J. Syst. Sci. Complex. **36**(5), 1789–1808 (2023)
32. Duan, G.R.: Substability and substabilization: control of subfully actuated systems. IEEE Trans. Cybern. **53**(11), 7309–7322 (2023)
33. Fan, C., Zhu, H., Chen, X., Wei, W., Gu, M., Long, X.: Initial analysis of anomalies in the first flight of spacex's super heavybooster and starship. J. Astronaut. **44**(5), 805–813 (2023)
34. Fei, Y., Shi, P., Lim, C.P., Yuan, X.: Finite-time observer-based formation tracking with application to omnidirectional robots. IEEE Trans. Ind. Electron. **70**(10), 10598–10606 (2023)
35. Freidovich, L.B., Khalil, H.K.: Performance recovery of feedback-linearization-based designs. IEEE Trans. Autom. Control **53**(10), 2324–2334 (2008)
36. Gao, C., Duan, G.: Fault diagnosis and fault tolerant control for nonlinear satellite attitude control systems. Aerospace Sci. Technol. **33**(1), 9–15 (2014)
37. Gao, Y.J., Duan, G.R.: Robust model reference tracking control for high-order descriptor linear systems subject to parameter uncertainties. IET Control Theory Appl. **18**(4), 479–494 (2024)
38. Gao, M., Niu, Y., Sheng, L.: Distributed fault-tolerant state estimation for a class of nonlinear systems over sensor networks with sensor faults and random link failures. IEEE Syst. J. **16**(4), 6328–6337 (2022)
39. Gauthier, J., Hammouri, H., Othman, S.: A simple observer for nonlinear systems applications to bioreactors. IEEE Trans. Autom. Control **37**(6), 875–880 (1992)

40. Ge, J., Sun, Y.: Analysis and Synthesis of Fault-Tolerant Control Systems. Zhejiang University Press, Hangzhou (1994)
41. Gitler, S.: Immersion and embedding of manifolds. In: Proceedings of Symposia in Pure Mathematics, vol. 22, pp. 87–96 (1971)
42. Gong, M., Sheng, L., Zhou, D.: Robust fault-tolerant stabilisation of uncertain high-order fully actuated systems with actuator faults. Int. J. Syst. Sci. **55**(12), 2518–2530 (2024)
43. Guo, B.Z., Zhao, Z.l.: On the convergence of an extended state observer for nonlinear systems with uncertainty. Syst. Control Lett. **60**(6), 420–430 (2011)
44. Guo, B.Z., Zhao, Z.L.: Weak convergence of nonlinear high-gain tracking differentiator. IEEE Trans. Autom. Control **58**(4), 1074–1080 (2012)
45. Guo, B.Z., Zhao, Z.L.: On convergence of the nonlinear active disturbance rejection control for MIMO systems. SIAM J. Control Optim. **51**(2), 1727–1757 (2013)
46. Hu, L., Duan, G., Hou, M.: Robust adaptive guaranteed cost tracking control for high-order nonlinear systems with uncertainties based on high-order fully actuated system approaches. Int. J. Robust Nonlinear Control **33**(13), 7583–7605 (2023)
47. Hu, L., Duan, G., Hou, M.: Robust switching adaptive tracking control for uncertain high-order fully actuated systems based on fully actuated system approaches. J. Franklin Inst. **361**(4), 106659 (2024)
48. Huang, C., Liu, Z., Chen, C.L.P., Zhang, Y.: Fast finite-time neuroadaptive consensus control for nonlinear nontriangular structured multiagent systems with uncertainty. IEEE Trans. Syst. Man Cybern. Syst. **53**(7), 4453–4465 (2023)
49. Huang, Q., Sun, J., Zhang, C.: High-order fully actuated system approach to robust control of impulsive systems. IEEE Trans. Circuits Syst. II Express Briefs **71**(3), 1321–1325 (2024)
50. Isidori, A.: Nonlinear Control Systems, 3rd edn. Springer, Berlin (1995)
51. Jia, F., Cao, F., He, X.: Active fault-tolerant control against intermittent faults for state-constrained nonlinear systems. IEEE Trans. Syst. Man Cybern. Syst. **54**(4), 2389–2401 (2024)
52. Jiang, J., Yu, X.: Fault-tolerant control systems: a comparative study between active and passive approaches. Ann. Rev. Control **36**(1), 60–72 (2012)
53. Khalil, H.K.: Extended high-gain observers as disturbance estimators. SICE J. Control Measur. Syst. Integr. **10**(3), 125–134 (2017)
54. Khosrowjerdi, M.J., Barzegary, S.: Fault tolerant control using virtual actuator for continuous-time Lipschitz nonlinear systems. Int. J. Robust Nonlinear Control **24**(16), 2597–2607 (2014)
55. Kim, J., Lee, C., Shim, H., Eun, Y., Seo, J.H.: Detection of sensor attack and resilient state estimation for uniformly observable nonlinear systems having redundant sensors. IEEE Trans. Autom. Control **64**(3), 1162–1169 (2019)
56. Lee, C., Shim, H., Eun, Y.: On redundant observability: from security index to attack detection and resilient state estimation. IEEE Trans. Autom. Control **64**(2), 775–782 (2019)
57. Li, P., Duan, G.: High-order fully actuated control approaches of flexible servo systems based on singular perturbation theory. IEEE/ASME Trans. Mechatron. **28**(6), 3386–3397 (2023)
58. Li, H., Shi, P., Yao, D., Wu, L.: Observer-based adaptive sliding mode control for nonlinear markovian jump systems. Automatica **64**, 133–142 (2016)
59. Li, Y., Sun, K., Tong, S.: Adaptive fuzzy robust fault-tolerant optimal control for nonlinear large-scale systems. IEEE Trans. Fuzzy Syst. **26**(5), 2899–2914 (2018)
60. Li, Y.X., Hou, Z., Che, W.W., Wu, Z.G.: Event-based design of finite-time adaptive control of uncertain nonlinear systems. IEEE Trans. Neural Netw. Learn. Syst. **33**(8), 3804–3813 (2022)
61. Li, X., Chen, Y., Sun, H.J., Liu, W.: Adaptive iterative learning control for high-order nonlinear systems with different types of uncertainties. Int. J. Robust Nonlinear Control **34**(8), 5399–5418 (2024)
62. Liang, B., Duan, G.: Robust $\mathscr{H}_\infty$ fault-tolerant control for uncertain descriptor systems by dynamical compensators. J. Control Theory Appl. **2**(3), 288–292 (2004)
63. Liu, G.P.: Coordination of networked nonlinear multi-agents using a high-order fully actuated predictive control strategy. IEEE/CAA J. Autom. Sin. **9**(4), 615–623 (2022)
64. Liu, C., Jiang, B.: $\mathscr{H}_\infty$ fault-tolerant control for time-varied actuator fault of nonlinear system. Int. J. Syst. Sci. **45**(12), 2447–2457 (2014)

65. Liu, M., Shi, P.: Sensor fault estimation and tolerant control for itô stochastic systems with a descriptor sliding mode approach. Automatica **49**(5), 1242–1250 (2013)
66. Liu, P., Chen, S., Zhao, Z.L.: On active disturbance rejection control for lower-triangular systems with mismatched nonlinear uncertainties and unknown time-varying control coefficients. Nonlinear Dyn. **106**, 2377–2400 (2021)
67. Liu, X., Chen, M., Sheng, L., Zhou, D.: Adaptive fault-tolerant control for nonlinear high-order fully-actuated systems. Neurocomputing **495**, 75–85 (2022)
68. Liu, X., Chen, M., Zhou, D., Sheng, L.: Fault-tolerant control of stochastic high-order fully actuated systems. IEEE Trans. Cybern. **54**(5), 3225–3238 (2024)
69. Lu, S., Zhou, G., Jingdong, S.: Study on operation system reliability of high-speed railway based on mmese model-a case of "7.23" yong-wen line accident. J. Safety Sci. Technol. **9**(3), 19–25 (2013)
70. Lu, S., Tsakalis, K., Chen, Y.: Development and application of a novel high-order fully actuated system approach—part I: 3-DOF quadrotor control. IEEE Control Syst. Lett. **7**, 1177–1182 (2022)
71. Lu, S., Tsakalis, K., Chen, Y.: Development and application of a novel high-order fully actuated system approach: part II. 6-DOF quadrotor control. In: 2023 American Control Conference (ACC), pp. 661–666 (2023)
72. Lu, M., Chen, X., Wu, F., Cao, X.: Attitude maneuver control of spacecraft based on second-order fully actuated system under attitude constraints. Acta Aeronaut. Astron. Sin. **45**(1), 628958 (2024)
73. Ma, H.J., Yang, G.H.: Adaptive fault tolerant control of cooperative heterogeneous systems with actuator faults and unreliable interconnections. IEEE Trans. Autom. Control **61**(11), 3240–3255 (2016)
74. Ma, J., Park, J.H., Xu, S.: Global adaptive control for uncertain nonlinear systems with sensor and actuator faults. IEEE Trans. Syst. Man Cybern. Syst. **51**(9), 5503–5510 (2021)
75. Ma, Y., Jiang, B., Wang, J., Gong, J.: Adaptive fault-tolerant formation control for heterogeneous UAVs-UGVs systems with multiple actuator faults. IEEE Trans. Aerospace Electron. Syst. **59**(5), 6705–6716 (2023)
76. Meng, R., Hua, C., Li, K., Ning, P.: Adaptive event-triggered control for uncertain high-order fully actuated system. IEEE Trans. Circuits Syst. II Express Briefs **69**(11), 4438–4442 (2022)
77. Meza-Aguilar, M., Loukianov, A.G., Rivera, J.: Sliding mode adaptive control for a class of nonlinear time-varying systems. Int. J. Robust Nonlinear Control **29**(3), 766–778 (2019)
78. Niederlinski, A.: A heuristic approach to the design of linear multivariable interacting control systems. Automatica **7**(6), 691–701 (1971)
79. Niu, Y., Sheng, L., Gao, M., Zhou, D.: Distributed intermittent fault detection for linear stochastic systems over sensor network. IEEE Trans. Cybern. **52**(9), 9208–9218 (2022)
80. Patton, R.: Robustness issues in fault-tolerant control. In: IEE Colloquium on Fault Diagnosis and Control System Reconfiguration, pp. 1–25. IET (1993)
81. Patton, R.J.: Fault-tolerant control: the 1997 situation. IFAC Proc. Vol. **30**(18), 1029–1051 (1997)
82. Ran, M., Wang, Q., Dong, C.: Active disturbance rejection control for uncertain nonaffine-in-control nonlinear systems. IEEE Trans. Autom. Control **62**(11), 5830–5836 (2017)
83. Ran, M., Wang, Q., Dong, C., Xie, L.: Active disturbance rejection control for uncertain time-delay nonlinear systems. Automatica **112**, 108692 (2020)
84. Ran, M., Li, J., Xie, L.: A new extended state observer for uncertain nonlinear systems. Automatica **131**, 109772 (2021)
85. Rosier, L.: Homogeneous lyapunov function for homogeneous continuous vector field. Syst. Control Lett. **19**(6), 467–473 (1992)
86. Seron, M.M., De Dona, J.A., Olaru, S.: Fault tolerant control allowing sensor healthy-to-faulty and faulty-to-healthy transitions. IEEE Trans. Autom. Control **57**(7), 1657–1669 (2012)
87. Sharida, A., Bayhan, S., Abu-Rub, H., Fesli, U.: Voltage-sensorless open-switch fault-tolerant control of three-phase T-type rectifier. IEEE Trans. Power Electron. **38**(12), 15365–15376 (2023)

88. Shim, H.: A passivity-based nonlinear observer and a semi-global separation principle. Ph.D. Thesis, Seoul National University (2000)
89. Siljak, D.D.: Reliable control using multiple control systems. Int. J. Control **31**(2), 303–329 (1980)
90. Stevens, B.L., Lewis, F.L.: Aircraft Control and Simulation. Wiley, Hoboken (1992)
91. Sui, S., Tong, S.: Finite-time fuzzy adaptive PPC for nonstrict-feedback nonlinear MIMO systems. IEEE Trans. Cybern. **53**(2), 732–742 (2023)
92. Tian, G., Duan, G.: Robust model reference tracking for uncertain second-order nonlinear systems with application to robot manipulator. Int. J. Robust Nonlinear Control **33**(3), 1750–1771 (2023)
93. Tian, G., Li, B., Zhao, Q., Duan, G.: High-precision trajectory tracking control for free-flying space manipulators with multiple constraints and system uncertainties. IEEE Trans. Aerospace Electron. Syst. **60**(1), 789–801 (2024)
94. Tu, Y., Wang, D., Ding, S.X., Fu, F., Li, W.: A reconfiguration-based fault-tolerant control method for nonlinear uncertain systems. IEEE Trans. Autom. Control **67**(11), 6060–6067 (2022)
95. Wang, L., Kellett, C.M.: Robust output feedback stabilization of MIMO invertible nonlinear systems with output-dependent multipliers. IEEE Trans. Autom. Control **67**(6), 2989–2996 (2022)
96. Wang, N., Liu, X., Liu, C., Wang, H., Zhou, Y.: Almost disturbance decoupling for HOFA nonlinear systems with strict-feedback form. J. Syst. Sci. Complexity **35**(2), 481–501 (2022)
97. Wang, Y., Chen, M., Lvwen, Z.: Dynamic simulation and analysis of "brake failure" of tesla electric vehicle. Mech. Eng. **44**(4), 852–856 (2022)
98. Wang, Y., Chen, Z., Sun, M., Sun, Q., Piao, M.: On sign-projected gradient flow-optimized extended-state observer design for a class of systems with uncertain control gain. IEEE Trans. Ind. Electron. **70**(1), 773–782 (2023)
99. Wang, L., Liu, P.X., Wang, H.: Fast finite-time control for nonaffine stochastic nonlinear systems against multiple actuator constraints via output feedback. IEEE Trans. Cybern. **53**(5), 3253–3262 (2023)
100. Wang, L., Liu, Y.X., Zhang, C.H., Hua, C.C., Wei, Y.Q.: Adaptive event-triggered control for the nonlinear cascade systems: a HOFA system approach. In: 2023 IEEE 13th International Conference on CYBER Technology in Automation, Control, and Intelligent Systems (CYBER), pp. 356–361. IEEE, Piscataway (2023)
101. Wang, N., Liu, X., Liu, C., Wang, H.: Weak disturbance decoupling of high-order fully actuated nonlinear systems. Int. J. Robust Nonlinear Control **34**(3), 1971–2012 (2024)
102. Warner, F.W.: Foundations of Differentiable Manifolds and Lie Groups. Springer, New York (1983)
103. Wu, Y., Liu, J., Wang, Z., Ju, Z.: Distributed resilient tracking of multiagent systems under actuator and sensor faults. IEEE Trans. Cybern. **53**(7), 4653–4664 (2023)
104. Xiao, Y., Cai, G., Duan, G.: High-order adaptive dynamic surface control for output-constrained non-linear systems based on fully actuated system approach. Int. J. Syst. Sci. **55**(3), 482–498 (2024)
105. Xie, C.H., Yang, G.H.: Data-based fault-tolerant control for affine nonlinear systems with actuator faults. ISA Trans. **64**, 285–292 (2016)
106. Xu, N., Chen, L., Yang, R., Zhu, Y.: Multi-controller-based fault tolerant control for systems with actuator and sensor failures: application to 2-body point absorber wave energy converter. J. Franklin Inst. **359**(12), 5919–5934 (2022)
107. Yan, S., Sun, W., Yu, X., Gao, H.: Adaptive sensor fault accommodation for vehicle active suspensions via partial measurement information. IEEE Trans. Cybern. **52**(11), 12290–12301 (2022)
108. Yan, F., Zhang, M., Gu, G.: Adaptive estimation and control for uncertain nonlinear systems and full actuation control. Sci. China Inf. Sci. **66**(11), 212204 (2023)
109. Yan, C., Xia, J., Park, J.H., Feng, J.e., Xie, X.: Fully-actuated system approach-based dynamic event-triggered control with guaranteed transient performance of flexible-joint robot: experiment. IEEE Trans. Circuits Syst. II Express Briefs **71**(8), 3775–3779 (2024)

110. Yang, H., Jiang, B., Cocquempot, V.: Supervisory fault-tolerant regulation for nonlinear systems. Nonlinear Anal. Real World Appl. **12**(2), 789–798 (2011)

111. Yang, H., Jiang, Y., Yin, S.: Adaptive fuzzy fault-tolerant control for markov jump systems with additive and multiplicative actuator faults. IEEE Trans. Fuzzy Syst. **29**(4), 772–785 (2021)

112. Ye, Y., Pan, R., Weisun, J.: Review and prospect of fault-tolerant technology for control systems. In: Proceedings of the Second Process Control Science Paper Report, pp. 49–61 (1988)

113. Yin, X., Wang, X.: Review, analysis and prospect of self-healing control system. Control Theory Appl. **38**, 1145–1158 (2021)

114. Yin, S., Yang, H., Kaynak, O.: Sliding mode observer-based FTC for markovian jump systems with actuator and sensor faults. IEEE Trans. Autom. Control **62**(7), 3551–3558 (2017)

115. Young, A., Cao, C., Patel, V., Hovakimyan, N., Lavretsky, E.: Adaptive control design methodology for nonlinear-in-control systems in aircraft applications. J. Guidance Control Dyn. **30**(6), 1770–1782 (2007)

116. Yu, X., Wang, T., Gao, H.: Adaptive neural fault-tolerant control for a class of strict-feedback nonlinear systems with actuator and sensor faults. Neurocomputing **380**, 87–94 (2020)

117. Yu, J., Shi, P., Liu, J., Lin, C.: Neuroadaptive finite-time control for nonlinear MIMO systems with input constraint. IEEE Trans. Cybern. **52**(7), 6676–6683 (2022)

118. Yu, Y., Liu, G.P., Huang, Y., Guerrero, J.M.: Coordinated predictive secondary control for DC microgrids based on high-order fully actuated system approaches. IEEE Trans. Smart Grid **15**(1), 19–33 (2024)

119. Yu, Y., Liu, G.P., Huang, Y., Shi, P.: Optimal cooperative secondary control for islanded DC microgrids via a fully actuated approach. IEEE/CAA J. Autom. Sin. **11**(2), 405–417 (2024)

120. Yuan, P., Ma, Y.: The crash of boeing 737-max airplane and its quality enlightenment. Qual. Reliabil. **01**, 64–66 (2021)

121. Zavala-Rio, A., Fantoni, I.: Global finite-time stability characterized through a local notion of homogeneity. IEEE Trans. Autom. Control **59**(2), 471–477 (2014)

122. Zhang, Y., Hua, C.: Composite learning finite-time control of robotic systems with output constraints. IEEE Trans. Ind. Electron. **70**(2), 1687–1695 (2023)

123. Zhang, D.W., Liu, G.P.: Predictive sliding-mode control of networked high-order fully actuated systems under random deception attacks. Sci. China Inf. Sci. **66**(9), 190204 (2023)

124. Zhang, J.X., Yang, G.H.: Robust adaptive fault-tolerant control for a class of unknown nonlinear systems. IEEE Trans. Ind. Electron. **64**(1), 585–594 (2016)

125. Zhang, J.X., Yang, G.H.: Fault-tolerant output-constrained control of unknown Euler–lagrange systems with prescribed tracking accuracy. Automatica **111**, 108606 (2020)

126. Zhang, D.W., Liu, G.P., Cao, L.: Coordinated control of high-order fully actuated multiagent systems and its application: a predictive control strategy. IEEE/ASME Trans. Mechatron. **27**(6), 4362–4372 (2022)

127. Zhang, D.W., Liu, G.P., Cao, L.: Proportional integral predictive control of high-order fully actuated networked multiagent systems with communication delays. IEEE Trans. Syst. Man Cybern. Syst. **53**(2), 801–812 (2022)

128. Zhang, H., Mu, Y., Gao, Z., Wang, W.: Observer-based fault reconstruction and fault-tolerant control for nonlinear systems subject to simultaneous actuator and sensor faults. IEEE Trans. Fuzzy Syst. **30**(8), 2971–2980 (2022)

129. Zhang, K., Jiang, B., Ding, S.X., Zhou, D.: Robust asymptotic fault estimation of discrete-time interconnected systems with sensor faults. IEEE Trans. Cybern. **52**(3), 1691–1700 (2022)

130. Zhang, L., Zhu, L., Hua, C.: Practical prescribed time control based on high-order fully actuated system approach for strong interconnected nonlinear systems. Nonlinear Dyn. **110**(4), 3535–3545 (2022)

131. Zhang, L., Wang, P., Hua, C.: Adaptive control of time-delay nonlinear HOFA systems with unmodeled dynamics and unknown dead-zone input. Int. J. Robust Nonlinear Control **33**(4), 2615–2628 (2023)

132. Zhang, H., Wang, W., Chen, S., Ji, Y., Liu, J.: Integrated guidance and control design based on fully actuated system method. Acta Aeronaut. Astronaut. Sin. **45**(1), 628891 (2024)

133. Zhao, Z.L., Guo, B.Z.: On convergence of nonlinear active disturbance rejection control for SISO nonlinear systems. J. Dyn. Control Syst. **22**(2), 385–412 (2016)

134. Zhao, Z.L., Jiang, Z.P.: Finite-time output feedback stabilization of lower-triangular nonlinear systems. Automatica **96**, 259–269 (2018)

135. Zhao, D., Polycarpou, M.M.: Distributed fault accommodation of multiple sensor faults for a class of nonlinear interconnected systems. IEEE Trans. Autom. Control **67**(4), 2092–2099 (2022)

136. Zhao, Z., Wang, Z., Zou, L., Guo, G.: Finite-time state estimation for delayed neural networks with redundant delayed channels. IEEE Trans. Syst. Man Cybern. Syst. **51**(1), 441–451 (2021)

137. Zhao, J., Lu, P., Du, C., Cao, F.: Active fault-tolerant strategy for flight vehicles: transfer learning-based fault diagnosis and fixed-time fault-tolerant control. IEEE Trans. Aerospace Electron. Syst. **60**(1), 1047–1059 (2024)

138. Zhao, Z., Zhang, J., Liu, Z., Li, H.X., Chen, C.P.: Event-triggered adaptive neural fault-tolerant control for a 2-dof helicopter system with prescribed performance. Automatica **162**, 111511 (2024)

139. Zhou, D., Ding, X.: Fault-tolerant control theory and its applications. Acta Autom. Sin. **26**(6), 788–797 (2000)

140. Zhou, D., Frank, P.: Fault diagnostics and fault tolerant control. IEEE Trans. Aerospace Electron. Syst. **34**(2), 420–427 (1998)

141. Zhou, D., Ye, Y.: Modern Fault Diagnosis and Fault-Tolerant Control. Tsinghua University Press, Beijing (2000)

142. Zou, Y., Xia, K.: Robust fault-tolerant control for underactuated takeoff and landing UAVs. IEEE Trans. Aerospace Electron. Syst. **56**(5), 3545–3555 (2020)